PLANT SPIRIT SHAMANISM

PLANT SPIRIT SHAMANISM

TRADITIONAL TECHNIQUES
FOR HEALING THE SOUL

ROSS HEAVEN
AND
HOWARD G. CHARING

FOREWORD BY PABLO AMARINGO

Destiny Books
Rochester, Vermont

Destiny Books
One Park Street
Rochester, Vermont 05767
www.DestinyBooks.com

Destiny Books is a division of Inner Traditions International

Library of Congress Cataloging-in-Publication Data
Heaven, Ross.
  Plant spirit Shamanism : traditional techniques for healing the soul / Ross Heaven and Howard G. Charing ; foreword by Pablo Amaringo.
     p. cm.
  Includes bibliographical references and index.
  ISBN-13: 978-1-59477-118-7 (pbk.)
  ISBN-10: 1-59477-118-9 (pbk.)
  1. Shamanism. 2. Plants—Miscellanea. 3. Spiritual healing. I. Charing, Howard G. II. Title.
  BF1623.P5H43 2006
  201'.44—dc22
                                                                    2006013312

Printed and bound in the United States by Lake Book Manufacturing

10 9 8 7 6 5 4

Text design and layout by Virginia Scott Bowman
This book was typeset in Sabon with Galliard as a display typeface

To send correspondence to the author of this book, mail a first-class letter to the author c/o Inner Traditions • Bear & Company, One Park Street, Rochester, VT 05767, and we will forward the communication.

For my children: Ocean, Jodie, and Mili. And with thanks to Donna, Andy, Leticia, Ram, and Howard, the case for the sane. Outcome inevitable and now a trail of truth back to the ocean.

R. H.

For my daughters, Katie and Elizabeth. And with appreciation and thanks to Peter Cloudsley for his assistance and support in preparing this book. Also to my friend and colleague Leo Rutherford, for his encouragement.

H. C.

We also thank the maestros we have worked and studied with and who are mentioned in this book: Artidoro Aro Cardenas, Guillermo Arevalo, Javier Arevalo, Doris Rivera Lenz, Juan Navarro, Loulou Prince, and our jungle guides and interpreters: Gerlo in Peru, and Jacqui in Haiti. We appreciate your time and energy in sharing ceremonies with us and in putting up with our endless questions and interviews.

DISCLAIMER

The journeys, plants, diets, and recipes offered in this book are for interest purposes only.

The exercises we present have been tested in many workshops, client healings, and other real-life applications. No harm has ever arisen for any participant as a result, and most have benefited enormously. It *is important,* however, to act sensibly and responsibly when using any plant medicines and to double-check all formulas and recipes for legality and safety before using them internally or externally. We also advise a medical consultation before embarking on plant diets, to reassure yourself that there are no contraindications to the plants listed herein.

Any application of these exercises is at the reader's own risk, and the authors and publishers disclaim any liability arising directly or indirectly from the use of this book, its exercises, or the plants and medicines it describes.

CONTENTS

THE PLANTS GAVE ME LIFE

I owe my life to plants, and they have informed everything I have done. From the time when I was very young, I liked to work with plants and knew that they gave me daily sustenance, not just as food for my body, but in my soul. I loved and admired them greatly. In my adolescence they became even more important to me. I was very unwell in my heart,* but I healed myself with the sacred plant, ayahuasca, after years of suffering—something that medicines from the pharmacy were unable to do.

After years of healing myself in this way, I saw a *curandera* (female shaman) heal my younger sister, also using ayahuasca. My sister had been in mortal agony with hepatitis, but with this single healing from the plants, she was cured in just two hours. That motivated me to start learning the science of vegetalismo.† Later I began dieting with la purga,‡

---

*Pablo Amaringo was born in 1943 in Puerto Libertad, in the Peruvian Amazon. He was ten years old when he first took ayahuasca—a visionary brew used in shamanism—to help him overcome a severe heart disease. The magical cure of this ailment via the healing plants led Pablo toward the life of a shaman.

†A vegetalismo is a shamanic healer who works primarily with plants.

‡La purga is another name for ayahuasca. It is given this name because of its powerful purging qualities, which are seen as a beneficial means of physical and spiritual cleansing.

and she taught me how to use plants for healing and how to understand their application through visions. That's how I came to be a shaman, ordained by the spirits.

My visions helped me understand the value of human beings, animals, the plants themselves, and many other things. The plants taught me the function they play in life, and the holistic meaning of *all* life. We all should give special attention and deference to Mother Nature. She deserves our love. And we should also show a healthy respect for her power!

Plants are essential in many ways: they give life to all beings on Earth by producing oxygen, which we need to be active; they create the enormous greenhouse that gives board and lodging to diverse but interrelated guests; they are teachers who show us the holistic importance of conserving life in its due form and necessary conditions.

For me personally, though, they mean even more than this. Plants—in the great living book of nature—have shown me how to study life as an artist and shaman. They can help all of us to know the art of healing and to discover our own creativity, because the beauty of nature moves people to show reverence, fascination, and respect for the extent to which the forests give shelter to our souls.

The consciousness of plants is a constant source of information for medicine, alimentation, and art, and an example of the intelligence and creative imagination of nature. Much of my education I owe to the intelligence of these great teachers. Thus I consider myself to be the "representative" of plants, and for this reason I assert that if they cut down the trees and burn what's left of the rainforests, it is the same as burning a whole library of books without ever having read them.

People who are not so dedicated to the study and experience of plants may not think this knowledge is so important to their lives—but even they should be conscious of the nutritional, medicinal, and scientific value of the plants they rely on for life.

My most sublime desire, though, is that every human being should begin to put as much attention as he or she can into the knowledge of plants, because they are the greatest healers of all. And all human beings should also put effort into the preservation and conservation of the rainforest and care for it and the ecosystem, because damage to these not only prejudices the flora and fauna but humanity itself.

Even in the Amazon these days, many see plants as only a resource for building houses and to finance large families. People who have farms and raise animals also clear the forest to produce foodstuffs. Mestizos and native Indians log the largest trees to sell to industrial sawmills for subsistence. They have never heard of the word ecology!

I, Pablo, say to everybody who lives in the Amazon and the other forests of the world, that they must love the plants of their land, and everything that is there!

This expression of love must be a sincere and altruistic interest in the lasting well-being of others. We are not here simply to exist, but to enjoy life together with plants, animals, and loved ones, and to delight in contemplation of the beauty of nature. A shaman has in his mind and heart the attitude of conserving nature because he knows that life is for enjoying the company of this world's countless delights.

Any painting or book or piece of art that spreads this message is to be respected, and every reader who picks up a book on this subject is to be honored.

I invite you to read on and to learn from the greatest teachers of all—the plants, our sacred brothers and sisters.

PABLO C. AMARINGO, OCTOBER 2005
(TRANSLATED BY PETER CLOUDSLEY)

Pablo Amaringo is one of the world's greatest visionary artists. A member of the Shipibo tribe, he trained as a *curandero* (shaman) in the Amazon, healing himself and others from the age of ten, but gave this up in 1977 to become a full-time painter and art teacher at his Usko-Ayar school. His book, *Ayahuasca Visions: The Religious Iconography of a Peruvian Shaman*, co-authored with Luis Eduardo Luna, was published in 1993 by North Atlantic Books. He has a Web site at www.pabloamaringo.com.

PREFACE

SINS, SOULS, AND SUN FLOWERS

DISCOVERING THE POWER OF THE PLANTS

By Ross Heaven

> *In the County of Hereford was an old Custom at Funerals, to hire poor people, who were to take upon them all the Sins of the party deceased . . .*
> *The manner was that when a Corpse was brought out of the house and laid on the Bier; a Loaf of bread was brought out and delivered to the Sin-eater over the corpse, as also a Mazer-bowl full of beer, which he was to drink up, and sixpence in money, in consideration whereof he took upon him all the Sins of the Defunct, and freed him (or her) from Walking after they were dead.*
> JOHN AUBREY, 1688

When I was a young boy, about nine years old, my family moved to the Herefordshire countryside, a place of mystery shadowed by the Welsh Black Mountains and the soul of Celtic myth.

At the edge of the village, alone and isolated from the rest of the small community, there was a cottage, long fallen to disrepair; it was a place I was warned to stay clear of. For in this cottage there lived a madman, unsafe and undesirable to the village. It was said that his house was haunted and a "place of strange lights." Of course, I found my way there almost immediately.

His cottage stood at a crossroads, just back from the road itself, and was surrounded by tall bushes and trees. It was a walk of about a mile from the village and there were no other houses near it. It looked and felt like the fairy-tale cottage of a witch, a place you stumble upon in error, after which your life cannot be the same.

As I stood looking at this mysterious cottage whose lopsided architecture had begun to take on the form of the land itself, the door opened and its single inhabitant emerged. He was old, dwarfed with age, and thin, dressed oddly for the times (the 1970s) in a white collarless shirt, black trousers, and waistcoat, like an old-fashioned country doctor or the cinema version of a period railroad signalman. A gold chain and fob watch hung from his waistcoat.

So this was the madman I had been warned away from. His name was Adam, and he began to talk to me about flowers and herbs and the story of his life. He had been a sin eater.

There is little written or known about sin eating, beyond that it was an ancient profession, once practiced in many countries of the world but now largely gone. As a healing tradition, it supposedly derived from the scapegoat ritual described in Leviticus, where the wrongdoings of the people were ascribed to an innocent. In the Biblical reference, Aaron confessed the sins of the children of Israel above the head of a goat that was then sent into the wilderness, carrying their wrongdoings and releasing them when it died.

In a similar way, a sin eater would be employed by the family of a deceased person, or sometimes by the church, to eat a last meal of bread and salt from the belly of the corpse as it lay in state. By so doing, it was believed, the sins of the dead would be absorbed by the sin eater himself and the deceased would have clear passage to the Kingdom of Heaven. The sin eater was given a few coins for his trouble, but other than that, he was avoided like the plague by a community that regarded him as sin filled and unclean because of his work. That is why sin eaters lived at the edge of the village and children were warned away from them.

The role of the sin eaters was, in essence, that of shamanic healers. Their job was to remove the spirit of disease from the dead (and often the living) and make the gods available to them. Their teachers in this work were the spirits themselves and the natural world.

Many sin eaters, because of their closeness to nature and rural location, were also skilled in folk medicine, "root doctoring," or herbalism, and they worked with both the medicinal properties and the spirit of the plants.

For example, vervain was known, medicinally, for its relaxing qualities and its ability to lift depression and restlessness. Adam, however, would interpret such symptoms as diseases of the soul, which were caused by guilt or by the shame of being in the presence of sin.* The spirit of vervain did more than heal the body, therefore; it eased the burden of the soul and drove away "evil spirits" (the effects, or the aura, of sin). By the same token, marigolds could be used to treat skin rashes, inflammation, and ulcers—but these were often also considered sin induced and symptoms of a deeper pain. Because marigolds were "bright like the sun" ("sun flowers," as Adam called them), they would soothe and uplift the soul.

A sin eater might also offer his patient advice from "the land of the dead" (the spirit world) for how his or her sins could be atoned for. This advice was often of a practical nature, the belief being that sins need to be reversed in this lifetime, not in the hereafter, and with action in the world, rather than prayers for forgiveness. The penitent might therefore be advised to make an offering, not to the spirits, but to the person he or she had wronged. The word *atonement* was key in this,

---

*In this tradition, guilt is regarded as a message from the soul that the patient has done something out of keeping with his or her own soul mission. Shame, meanwhile, is the imposition from others of their own guilt onto the patient.

such recompense being not a punishment but a means of rebalancing the soul so the patient could come back to *at-one-ment*.

In this way, sin eating—a practice thousands of years old—anticipated a mind-body-spirit connection that modern science is only just starting to notice, for its rituals and plant medicines worked not only on the patient's body, but also on the mind and emotions—and always on the troubled soul.

The introduction to this new realm was a fascinating adventure for a young boy and, as a result, I developed an enduring interest in plants and the spiritual, psychological, and emotional reasons for their use. When I grew up and began to travel, I explored other indigenous ways of working with plants. What I found among traditional healers was a remarkable consistency of approach. Whether in Greece, Turkey, North America, Haiti, Ireland, or Peru, wherever plant work was done by a healer rather than a medical doctor, it was the spirit of the plants that mattered and not their chemical properties.

Of course, the plants themselves varied (but then, not always—*rosa sisa,* or marigold, is used in the Amazon, for example, to "call back the soul" in the same way Adam had used these "sun flowers" in his quiet Hereford village) but the approach remained the same: "The plants tell us how to use them," these healers would always say.

And how do they do this? "They talk to us in dreams, in visions, and in journeys." This is a totally different way of working from the way a medical doctor might go to some dry book to look up the effects of a synthetic medicine derived from a plant, without once going near the plant itself—perhaps even without realizing that the pills he or she gave out were developed in a lab to synthesize the actions of a plant.

The exact means of using plant magic might also vary among healers—in Haiti, as we will see, the shaman uses a pouch of herbs called a *paket* to draw out illness from the body of a patient, while in the Amazon the shaman uses a bundle of leaves called a *chacapa* to stroke or gently beat a patient and achieve the same result—but these are cultural nuances. What is more amazing is that healers on different continents, who have never heard of each other, practice much the same methods—and they do so, not because they've read it in a book, but because the spirit of the plants taught them how.

Eventually, I came across the work of another writer who had explored the world of plant spirit medicine, Elliot Cowan. His observations of Mexican Huichol practices were the same: "The spirit of that plant must come to you. It [plant medicine] won't work for you unless the spirit of the plant tells you how to prepare it and what it will cure."[1] It is always the spirit that heals.

My interest, then, is the cross-cultural use of plants and the intriguing possibility of distilling all of these approaches to their essential core to discover how this healing works. I am sure there is much more to learn, but the message I have received so far, in its simplest form, is this: *Listen to the plants and they will teach you all you need to know.* This is good advice, and we will practice it throughout this book.

THE CALL OF THE PLANTS

By Howard G. Charing

We are not talking about passive agents of
transformation; we are talking about an intelligence,
a consciousness, an alive and other mind, a spirit. . . .
Nature is alive and is talking to us.
This is not a metaphor.

TERRENCE MCKENNA

My first real encounter with the plant world occurred when I first arrived in the Amazon ten years ago. The moment I stepped off the airplane in Iquitos, I felt as if I had been hit by a bolt of energy. I felt so energized that I didn't sleep for the next two days: my senses were at a heightened state of awareness, and it was as if I could hear the heartbeat of the rainforest itself.

Iquitos is a city in the Amazon rainforest. There are no completed roads into it, and the only way to get there is via airplane or riverboat. In the nineteenth century it was the center of the rubber industry, but by the early twentieth century the rubber trade had moved to the Far East and the city had fallen into neglect and disrepair. It is now a place without apparent purpose, resplendent in postcolonial splendor, but literally in the middle of nowhere—a true frontier town.

I recall my first moments in Iquitos: standing on the *malecon* (promenade) at the edge of the city, overlooking the river with some three thousand miles of pure rainforest spread out beyond me, I felt an exhilaration that still fills me with wonder and awe. I had come to Iquitos out of a longstanding interest and desire to experience at firsthand the living tradition of plant spirit medicines and of course the magical *ayahuasca* I had heard so much about. I was not to be disappointed; my first ayahuasca sessions with a shaman in a jungle clearing brought about a spiritual epiphany that changed my view and understanding of life. I experienced being at the very center of creation, with the realization and experience that I was not in any way separate, but an intrinsic part of the vast cosmic mind or field of consciousness, and that we were all connected, all part of this one great mind, our experience of separateness being no more than an illusion generated by our senses.

My ayahuasca sessions showed me much about the great dream of the Earth and how disconnected we humans had become from our relationship with the living planet. I experienced the evolutionary process of DNA and was shown that many of the problems that we as a species experience occur because we are basically primates, driven by our monkey glands and hormonal systems; yet at the same time we are also in harmony with a higher consciousness that approaches that of dolphins, which I experienced during the ayahuasca ceremonies as being part of a group consciousness.

In one of my most profound experiences, I found myself transported to what I perceived as the center of creation, where I witnessed planets, stars, nebulae, and universes being formed. Everywhere stretched vast patterns of intricate geometric and fluid complexity, constantly changing in size and form. The chanting of the shaman was filling every cell with an electric force; every part of my body was vibrating, and I felt as if I was being lifted into the air. I was in a temple of sound, vibration, and bliss. Gathered around me were giants in ornate costumes of gold and multicolored feathers blowing smoke and fanning me. These were the spirits of ayahuasca, whose soft, gentle, and exquisitely sensual voices spoke to me of creation and the universal mind. To reinforce this poetic insight, their words appeared before me in bold neon script.

Many of my visionary experiences with ayahuasca were personal, leading to a deeper understanding of my life and the role that various people had played in it. Sometimes I *became* those people, lived their lives, and came to understand why they did what they did, what decisions they had had to make in their lives. These experiences invariably led to either some form of closure, as if an open chapter had been completed, or a profound healing of my relationship with that person.

An example of this pertained to my mother. We had had a difficult relationship, a part of me never fully trusting her. During one session, I experienced a deep visionary communion with the Earth Mother; I understood the love that the spirit of our planet has for all who live within her. The vision and the experience of being gently held in a loving embrace was exquisite and sublime. Then the vision changed to my own mother—and I relived her life, her childhood, the things that had happened to her, the reason why she was taken away from her family home, the decisions that she'd had to make to have a child. And then I realized that I had judged her, and it was her misfortunes that had created this lack of trust between us.

After the session I was filled with joy, knowing that the gulf between us had been bridged; and ever since then the affection between us has flowed freely. That wonderful insight I owe completely to ayahuasca, the exquisite gift of the plant world.

I discovered that ayahuasca is a medicine, but one so unlike the Western understanding of medicine, which applies mainly to healing of physical symptoms. Ayahuasca is a medicine that works on every level: on the physical and nonphysical being, on our consciousness, on our emotions, and on our spirit. It is as if we are imbibing not just a liquid brew, but an "other" intelligence that knows exactly what is needed to help us. This act constitutes a communion in the true sense of the word. It can be an intense experience of euphoria: a journey of deep and profoundly meaningful personal and transpersonal insights. It can shine a searchlight on the hidden thoughts and feelings in the subconscious mind. It can bring an erasing of the ego boundaries and a merging with the whole of consciousness and creation.

There was more and more to experience. With some of the shaman's other teacher plants, such as *ajo sacha*, I felt my senses being altered—expanded in some ineffable way—and I became aware of the song of the rainforest. There were sounds, smells, and sights all around me that I had not been aware of in my normal everyday waking state. I could zoom in on the scents and the sounds, knowing that the rainforest is one entity, with the insects, birds, and animals as parts of its totality. I was in paradise lying in my hammock there, floating in a living, three-dimensional experience of sound, color, smell, movement, and vibration, all in harmony and great beauty.

Since that first encounter, I have spent many months in the rainforest working with and learning from the shamans, the holders of this beautiful knowledge. My enthusiasm and interest has never flagged. Plant knowledge has helped my life become vibrant and vital in so many ways.

THE WAY OF
THE PLANT SHAMAN

The first act of awe, when man was struck
with the beauty or wonder of Nature,
was the first spiritual experience.
HENRYK SKOLIMOWSKI

Since the beginning of human experience, plants have been partners in
the evolution of our species, not only by the provision of food and medicine but in our spiritual experience and developing consciousness. The
form, beauty, enchanting scents of the plant world, its healing and spiritual qualities, all provide a gateway to the Great Mystery of nature that
our Celtic forebears called "the visible face of spirit."

Though our lands are no longer forested as they once were, we try to
re-create this beauty and tranquility with our flower-filled gardens and
parks. The green spaces within our modern cities give us at least a taste
of nature with which we can sustain ourselves against the soulless back-
drop of the steel and concrete jungles that most of us now call home.

For many people, plants are still the messengers of divinity, harmony,
and beauty. They are also the source of our health and well-being, not
just as medicines but by their innate ability to relax, refresh, or excite us.
Many readers will remember making a home in the arms of great trees

1

in the treehouses we built as kids, for example, or lying in lush meadow grass gazing up at a clear blue sky on some perfect summer's day, and even the memories of these things can nourish us. They tell us there is still a possibility for poetry in our souls, no matter how mundane our lives have become.

Some deep part of us knows that the healing power is inherent in what plants are as much as what they do. We give flowers to friends who are unwell, for example, understanding in our souls, if not our conscious minds, that such gifts are uplifting and affirming of life. Flowers have a role to play, in fact, in all of our most primal celebrations of life and death—birth and birthdays, comings of age, marriages, illnesses, funerals. They are there for the beginnings of things, at the first "I love you," and they are there for our endings too. Even after death our connection to the natural world continues, as our ultimate spiritual destination within many religious myths is some form of paradise often symbolized as the "heavenly garden," or the Garden of Eden.*

Our relationship with the plant world is embedded within our consciousness, but since the advent of our materialistic civilization, we are often no longer aware of the mythic and spiritual connections to nature that we feel. In our synthetic world where, for example, we treat diseases with laboratory-developed chemicals, many people literally do not know that most of these drugs owe their existence and effectiveness, not to scientists in white coats, but to rainforest shamans and the plants that they work with. We have lost our connection to, and understanding of, nature—all that it gives us and all that we owe it—because, instead of remaining a part of nature, human beings have chosen to redefine themselves as the "masters" of nature.

---

*Of passing interest here is that the word *paradise* actually stems from the Indo-European, *pairidaçza*, meaning "a wall enclosing a garden or orchard." Xenophon, a Greek mercenary who spent time in the Persian army and later wrote histories and memoirs, uses the word *paradeisos*, but in his descriptions it denotes, not a wall, but the huge parks that Persian nobles loved to build and hunt in. It is Xenophon's word that is used in the Septuagint translation of Genesis to refer to the Garden of Eden, and from where Old English eventually borrowed it around 1200 C.E. And that is how a wall became a garden!

MASTERS OF NATURE?

The earth does not belong to man;
man belongs to the earth.

CHIEF SEATHL

This redefinition of ourselves as masters of the Earth has been most pronounced in the three hundred or so years since the start of the industrial
revolution, but in fact goes back many thousands of years, to a time
when we ceased to be nomads and became agriculturists, remaining in
one place and harnessing nature to our needs.

The historical fact is captured in the mythology of the Bible where,
in the book of Genesis, we read that we were "given dominion over all
living things." Some anthropologists argue that this reference was, in
fact, a form of agricultural public relations, as the settlers fought a war
for land with the nomadic tribes of the early Middle East. For these
anthropologists, the Fall, when humankind was evicted from Eden, is
symbolic of agricultural victory. The nomads, once an intrinsic part of
nature, were no longer free to roam their gardens, for now the gates
were guarded.*

Our separation from nature reached another critical point in the
sixteenth century, with the birth of modern science and the work of
the French philosopher, Rene Descartes. When he was a young soldier
in the Hapsburg army at the siege of Ulm, Descartes had a vision in
which a supernatural being (a giant winged angel) appeared to him and
told him that "nature can be conquered by measurement." This message
transformed his life and led him to develop the rationalist philosophy
that came to be known as Cartesian dualism. This is the belief that body
and mind are separate and that human beings, and all natural phenomena, are merely physical entities. In this philosophy, the soul—which
cannot be "proved" to exist, and so doesn't—is quite unnecessary. And
from this all science stems.

---

*For a novelist's interpretation of this anthropological evidence, see Daniel Quin, *Ish-
mael: An Adventure of the Mind and Spirit* (Bantam, 1995).

Of course, there is an incredible irony in this—that material science was founded on the words of an *immaterial* being. Nevertheless, it was the message of measurement that stuck and not the spiritual source of the words.

Where this denial of the soul, this separation from nature, and this fixation on measurement has led us is apparent. An estimated 100,000 chemicals are poured into our environment every day by companies operating within entirely scientific principles. Even our foods are not safe. Since we no longer trust nature to provide and care for us, millions of tons of pesticides are used on our crops each year. Millions more tons of toxins are released into the atmosphere through our food production techniques. We know that these are poisoning our immune systems because of the growing incidence of diseases such as asthma, irritable bowel syndrome, food allergies, yeast infections, Crohn's disease, colitis, arthritis, anxiety, insomnia, and depression—and yet we seem unable to stop killing ourselves, using processes that are destroying our rivers, air, and land. As long as we can "measure" the outcomes of pollution, the pollution itself has almost ceased to matter.

To give just one example, the 109 wineries in California's San Joaquin Valley have been calculated to produce 788 tons of smog-forming gases a year. This, according to one vineyard chief, is from an industry that "in general is for clean air. We are environmentally conscious." When industry decision makers are asked to clean up their act, however, environmental consciousness often takes a backseat to profits. "The problem here is that this is going to cost millions of dollars, and it's not even proven to work. And there would not even be that much of a benefit, because we really are not gross polluters."[1] How gross do you need to be? Apparently, 788 tons of smog per year is not gross enough.

As garden philosopher Masanobu Fukuoka puts it:

> Suppose that a scientist wants to understand nature. He may begin by studying a leaf, but as his investigation progresses, down to the level of molecules, atoms, and the elementary particles, he loses sight of the original leaf. . . . Which is to say that

research attempts to find meaning in something from which it has wrested all meaning.[2]

In other words, all the measurement in the world will bring us no closer to understanding the true nature of the world, or tell us how to behave responsibly, spiritually, and properly within it. Only we can do that.

Perhaps we need to differentiate between *fact* and *truth*. Our scientists can measure the facts of the natural world—that our rivers are ten times more polluted today than they were thirty years ago—but they cannot measure the deeper truths of the human condition and what it means to be a part of this polluted world. Nor can science tell us what to do about it. Only people—in touch with their souls—can know the truth.

PUTTING SOUL INTO SCIENCE

What I see in Nature is a magnificent structure that we can comprehend only very imperfectly, and that must fill a thinking person with a feeling of humility. This is a genuinely religious feeling.

ALBERT EINSTEIN

In our cultural imagination and general view of the world, we are still part of an archaic science based on measurement as the determining factor for how we should live: as masters of nature and separate from it. Modern scientists, however, now believe the very opposite of this archaic view that still permeates our culture. They have broken ranks with old *science* and returned to old *ways*, where the world is not just material but "soul-filled," alive, intelligent, and aware. The circle is almost complete as the serpent turns to eat its tail.

The current scientific model is quantum theory, which opens up a very curious universe indeed—one in which nothing *can* actually be measured, since the very action of doing so changes its nature, and the observer becomes the observed. In this universe, matter is made of particles and waves at one and the same time. To observe them, however, they can never be both. To measure (or even perceive) something, we must arrest the wave in space and time so that it becomes a particle. The *wave* we were looking at and aiming to measure then ceases to exist. It becomes something else: a wave stopped in time. Or we can continue to regard it as a wave, flowing, changing, and transient—but then it cannot be measured at all and, in fact, ceases to be *physical* as we understand the term, instead becoming energy in motion (or *e-motion,* a term that includes our feelings about it).

However we look at it, this "thing" we are observing does not exist in its own right. It is the choices we make and our behavior as observers that determines its reality, and even then, how we see it changes it. All we can really say is, "Things are, if we choose to believe they are," which begins to sound more like metaphysics than physics.

To make sense of the world at all, our scientists are, furthermore, now forced to talk about a reality founded on the notion of ten-dimensional hyperspace—the idea that there are other dimensions or invisible layers beyond the world we know. There is no proof for this, of course; but only by this supposition does reality become feasible. Without it, the world as we think we know it could not exist. But by believing in something they have no evidence for, our scientists have, in fact become theologians. Their argument that "one day we will find these other dimensions and prove that the world is as we 'know' it to be" is about as defensible as the argument of a priest that "one day we will find God and prove that the world is as we 'know' it to be."

What all of this amounts to is this: beyond the physical there is an *energetic* universe and we create it in our image every second of the day. This is no different than the shaman's claim that the world is made of spirit, not matter; or that "the world is as you dream it," as the Shuar people of the Amazon say. By "observing" a spiritual universe you *create* a spiritual universe; by seeing only the material as "real," that, also, is what you get.

For many shamans, this act of world creation relies on the spirits of the plants, which provide a gateway to the energy beyond the physical, and to the cosmos as a whole: the oneness of creation that we are part of, not standing back from as "caretakers," "conquerors," or "measurers."

Coming to terms with this concept can be a challenge for the rational Western mind. The proposition that a plant has intelligence or consciousness and we can communicate with it is something many regard as preposterous. For one thing, it requires an acceptance of the idea that plants can "speak," not only among themselves but with animals and humans as well.* This is not an easy idea for us to understand or accept. It requires a leap of imagination for us to open to the possibility of a direct encounter with the "other" consciousness or spirit of a plant.

Another challenge to our rational minds is that the entrance to this magical world is mainly through dream language or an expansion of the senses. This is less of a problem for non-Western cultures. For example, Loulou Prince, a *medsen fey* (leaf doctor and shaman-healer) in Jacmel, Haiti, says this of his own work with plants:

> I receive a lot of my knowledge in dreams. If I am treating a sick person, I often ask for a dream where I see the leaves I should give that person. In these dreams, the spirits come to me and tell me what to do, or I see that I am in the woods, and leaves are pushing up in front of me and these are the ones I should pick. Once I have this knowledge, I can make a remedy for the person who is suffering.

We in the West have tended to belittle our dreams. Our educational, political, economic, and legal systems are all based on measurement instead. Through these, we have become unused to working with our intuitive and imaginative selves and are often dismissive of them, regarding them as unimportant or in some way "second class."

Nevertheless, dreams and visions are our doorways to plant consciousness, and if we stop to think about it, this is not such a stretch after all. Every great achievement—whether a tall building, a child's birth, or a breakthrough discovery, such as Descartes' invention of rational

---

*As to whether plants can speak or not, see chapter 1 and make up your own mind.

science—begins first with an idea, a dream, or the revelations of the spirit. It does not just suddenly appear in the world. "Imagination," said Einstein, "is more important than knowledge."

THE MAGIC OF THE PLANTS

Despite the challenges for the Western mindset, there has been a huge surge of interest in plant spirit medicines in recent years, and in the knowledge of these plants that indigenous peoples hold. There is a growing belief that plants comprise a kind of "medicine for our times," offering new possibilities to people whom science has failed and providing inspirational solutions to the problems of civilization. In this book we look at the wisdom of our plant allies and show you how to work with them to develop your own communion with these great healers.

In chapter 1 we look at the range and scope of how plants heal. *Healing* is a concept that, in many cultures, goes beyond what Western culture normally regard as medicine (i.e., the administration of "physical cure" for a "physical illness") into the world of the spirit, where we can hear the underlying message of our disease and heal it through understanding, divination, the blessings of the plants, a change of luck, and by entering new realities so we experience a world beyond the physical, where all things can be known and there are still great frontier adventures left. Chapter 1 also explains, in practical terms, the shamanic principles underlying these different approaches so you can use them to make spirit allies of your own.

In chapter 2 we look at the concept of the shaman's diet, a body of practices that helps the shaman incorporate the plant spirit into his or her own. From this union, the plant itself informs and teaches the apprentice how to invoke its power so it can be used in healing. We also introduce some of the plants that shamans consider most important to develop a relationship with, such as *ajo sacha* and *chiric sanango*. All of these have psychospiritual as well as physical qualities that, interestingly, are able to adapt themselves to the needs of the culture and the person who diets them (i.e., ingests them in a disciplined and sustained way, with spiritual intentions). Thus, in the Amazon, where the diet is central to the training and initiation of the shaman, ajo sacha is used for hunting (to disguise the smell of the hunter and help him focus on his prey); in Europe and North America, however, this plant will help the dieter stalk (hunt down) more psychological, inner issues. The quality remains the same but the healing takes the form that is required.

Of course, knowing what happens in the Amazon or Haiti and what plants are available there is of limited value for a Western apprentice. For this reason, chapter 2 offers suggestions for more readily available plants or plant mixtures that will help you achieve the same effects, so you can heal yourself. (Also see appendices 1 and 2.)

Chapter 3 considers sacred hallucinogens. One of the ways that plants have always been used is (in Aldous Huxley's words) to "open the doors of perception" into new and other realities. This is a form of initiation into self; new neural pathways open up through communion with the plant, and the dieter may gain unusual and important insights into the nature of his—and all—life. In this chapter, we look at the ceremonial use of ayahuasca in the traditions of the Amazon, as well as the San Pedro ceremonies of the Andes. We also discover how to work with the essence of the universe opened up by such plants, in a way that does not itself involve the ingestion of hallucinogens.

Chapter 4 introduces two important concepts in shamanic healing—those of soul retrieval and spirit extraction. Though little documented before now, through these two practices the world over, plants are used to return vital energy and remove negative influences. This chapter explains soul retrieval and spirit extraction and demonstrates how plants are our allies in this. Often this healing work does not require the ingestion of plants at all, but their use in other ways, such as in the Mexican "ritual of flowers," the use of *pakets* in Haiti, which contain plants for sucking out spiritual toxins, the Amazonian use of *chacapas* (rattlelike bundles of leaves) to restore balance, and the Welsh sin-eating ritual of stinging with nettles to restore life and vitality. You will read how to make some of these medicine tools for yourself and how to offer a simple ceremony for the return of lost power.

Beautiful fragrances derived from flowers and herbs have long been used for healing and for visionary dreams—from the temple maidens of Egypt who danced with cones of incense to the floral baths of Haiti and Peru. Indeed, the word *perfume* originates from the Latin *per fumer,* "through smoke," a reference to ritual incense.

However, as we see in chapter 5, what has not been widely reported is that flower aromas can be used to change physical reality by altering one's "luck"—regarded as a real and controllable force in shamanic tradition. In the Amazon, specialists in this form of magic are known as *perfumeros,* and they are able to use smell, not just for adornment or in ritual, but for highly practical purposes, such as succeeding in business, winning court cases, and bringing back lovers who have strayed. There are parallels here to the Hoodoo tradition of New Orleans, which uses essential oils to similar effect.

These techniques are explained in chapter 5, and one of the best-kept (and most controversial) secrets of plant spirit medicine is also revealed: how the *pusanga* (the "love medicine of the Amazon") really works. We teach you how to find and make your own pusanga using herbs and flowers growing locally to you and provide recipes for perfumed oils to change your luck in love, money, health, and magic.

Chapter 6 explores floral baths as another way of removing energy blockages and revitalizing the spirit. These, too, are used in numerous shamanic traditions, including those of Brazil, Haiti, Indonesia, and Peru. They are also alluded to in Celtic legend, which informs the old sin-eating practices and the myth of the Grail. Many such baths exist— from flaming *kleren* baths, where a person bathes in fire, to relaxing perfumed baths, which cleanse the spirit and draw in love. In all cases, they work on our energies to restore balance and harmony.

Chapter 6 explains the energy body, how and why it becomes "blocked" and unbalanced, and what you can do to restore and re-energize yourself. It also includes recipes for baths you can treat yourself to, using easily found herbs and oils.

Finally, having introduced some of the main concepts and ways of working with the plant spirits, chapter 7 offers suggestions for how you might expand your work with these allies. It explains, for example, how you might conduct a healing in a modern setting, so you can pass on your knowledge in the treatment of others.

In the appendices at the end of the book you will find charts of Caribbean and Peruvian herbs, listing the commonly used plants of each culture, their healing and magical uses, and analogues that will allow you to continue your explorations into the world of plant spirits. Also included are Hoodoo recipes for luck and success. Following the appendices is a glossary of terms used throughout the book.

1

Nothing Is Hidden: How Plants Heal

*Since nothing is so secret or hidden that it cannot
be revealed, everything depends on the discovery
of those things which manifest the hidden. . . . He
who wishes to explore Nature must
tread her books with his feet.*

PARACELSUS

One concept that underlies all work with plants is that *nature itself will tell you what a plant is used for.* And nature's well-stocked medicine cabinet is right in front of you every day. Shamans distinguish the spiritual powers and qualities of plants in many ways: by the colors of their flowers, their perfumes, the shape and form of their leaves, where they are growing and in what ways, the moods they evoke, and the wider geographical, cultural, or mythological landscapes they occupy.

Although such considerations do not play a role in modern Western medicine (which does not believe in these spiritual powers at all), it was not long ago that we, too, had an understanding that nature is alive and talks to us in these ways. The sixteenth-century alchemist and philosopher Aureolus Phillippus Theophrastus Bombast—better known as Paracelsus—introduced this notion in his *Doctrine of Signatures* treatise, which proposed that the Creator has placed his seal on plants to indicate their medicinal uses.[1] This was not just idle speculation on the part of Paracelsus; nature itself taught him the truth of it.

"Seeking for truth," he wrote, "I considered within myself that if there were no teachers of medicine in this world, how would I set to learn the art? Not otherwise than in the great book of nature, written with the finger of God. . . . The light of nature, and no apothecary's lamp directed me on my way."

In this "book of nature," Paracelsus noted how the qualities of plants so often reflect their appearance—that the seeds of skullcap, for example, resemble small skulls and, it transpires, are effective at curing headache. Similarly, the hollow stalk of garlic resembles the windpipe and is used for throat and bronchial problems. And willow, which grows in damp places, will heal rheumatic conditions caused by a buildup of fluid on the joints.

In fact, as Thomas Bartram remarks in his *Encyclopedia of Herbal Medicine*, "Examples are numerous. It is a curiosity that many liver remedies have yellow flowers, those for the nerves (blue), for the spleen (orange), for the bones (white). Serpentaria (Rauwolfia) resembles a snake and is an old traditional remedy for snake-bite. Herbalism confirms the Doctrine of Signatures."[2]

Underlying Paracelsus's treatise was the premise that nature is itself a living organism that must be considered an expression of the "One Life," and that human and universe are the same in their essential nature. This idea was echoed (some would say proved) by Dr. James Lovelock, five hundred years after Paracelsus, in his Gaia hypothesis on the unity of life.[3] Lovelock shows that the Earth maintains relatively constant conditions in temperature and atmosphere to a degree that defies rational observations and predictive measurements of what "should" happen. It is, rather, as if the Earth is a living organism that purposefully takes care of itself, with human beings as part of that process. "We are, through our intelligence and communication, the nervous system of the planet," he wrote. "Through us, Gaia has seen herself from space, and begins to know her place in the universe. We should be the heart and mind of the

Earth, not its malady. . . . Most of all, we should remember that we are a part of it, and it is indeed our home."[4]

Because of his own farsighted belief in the planet as an aware and living organism, Paracelsus held that the inner nature of plants may be discovered by their outer forms or "signatures." He applied this principle to food as well as medicine, remarking that "it is not in the quantity of food but in its quality that resides the Spirit of Life"—a belief familiar to those who choose to eat organic food and share a common concern over genetically modified (GM) substitutes that lack life force, or spirit. According to Paracelsus, then, the appearance of a plant is the gateway to its spirit or consciousness.

The doctrine of signatures, per se, is not something known to many indigenous shamans, but they understand the principle behind it well enough—that nature is alive, aware, and communicates with us. This principle is not regarded as fanciful at all, but practical and important enough to save lives.

We discovered how the doctrine of signatures operates in the Amazon, for example, during an experience with the *jergon sacha* plant. Howard relates:

> We came across this plant quite accidentally one day while walking through the rainforest with the shaman Javier Arevalo, studying the properties of the plants. Javier queried why I always walked around with a machete. I jokingly replied, "It's against anacondas!"
>
> He paused for a moment, then beckoned me to follow him. A few minutes later we came across this tall-stemmed plant. This was jergon sacha, he said. Javier cut a stem from it and proceeded to whip me around the body, paying most attention to my legs and the soles of my feet. He then said, "No more problems, you are protected against snakes." I asked him why this plant was used in this way, and he indicated the pattern on the stem, which looks identical to the snakes in the forest. (See photos of jergon sacha on page 7 of the color insert.)
>
> Later, on a hunch, we started to investigate this plant and dis-

covered some amazing correspondences. Jergon sacha is widely used as an antidote to snake venom in the Amazon. Referring back to the concept of signatures, this plant is a clear demonstration of the outer form indicating the inner qualities. Its use is directly related to its physical appearance, the tall stem closely resembling the venomous pit viper known as the *jararaca*, or bushmaster, which is indigenous to the Amazon. The bushmaster, unlike most other snakes, is aggressive and will defend its territory. It can strike in the blink of an eye from fifteen feet and is rightly feared and respected.

Remarkably, jergon sacha does turn out to be a highly effective antidote for the bite when its large root tuber is chopped up and immersed in cold water and then drunk, or placed in a banana leaf and used as a poultice wrapped around the wound. Of course, pragmatically speaking, it is impossible to store antivenom vaccines in the rainforest, where there is no refrigeration, so the living plant has exceptional lifesaving importance. And its importance is recognized because the plant itself tells the shaman of its use through the markings on its stems.

Another illustration of the connection between the form and function of a plant is provided by Artidoro Aro Cardenas, a shaman who works with plant perfumes. "If the smell of a flower has the power to attract insects or birds, it can also attract luck to people," he says. Artidoro makes fragrances that attract customers into a shop, for example ("You just rub the perfume on your face and it brings in the people to your business"), as well as perfumes for love, and others for "flourishing"—growth and success. "I watch what the plant does and if it is attractive,* I use it to attract. Plants are the forces of nature," he says. "All I do is give these forces direction." (There is much more on the use of perfumes in chapter 5.)

The system of homeopathy is also based on the principle of a sentient universe known through its signatures. Hippocrates spoke of a universal law of *similia similibus curentur* (like cures like), and the modern pioneer of homeopathy, Samuel Hahnemann (1755–1843), showed, through his

---

* "Attractive"—that is, has the power to attract.

experiments, that plants contain a healing essence or spiritual quality that has an affinity with human beings and acts on them according to the nature of the illness from which they are suffering.

No one really knows how homeopathy works (by *no one,* here, we mean clinicians and scientists who do not believe in a nonphysical world); but the fact that it does seems clear. In 1836, for example, when cholera destroyed many Austrian cities and orthodox medicine was unable to stop its spread, the government turned in desperation to homeopathy and built a quick and crude hospital in which patients could be treated. The results spoke for themselves: while orthodox hospitals reported deaths in more than 70 percent of cases, the homeopathic hospital recorded a death rate of just 30 percent. Shamans have a simple explanation for this: the homeopathic doctors appealed to and engaged the spirit of the plants to intervene on behalf of their patients, and the spirits answered their call.

SYMPATHETIC MAGIC: LIKE PRODUCES LIKE, ALL THINGS ARE CONNECTED

The principles of magic, which operate throughout the shamanic world and were identified by Sir James Frazer in his book *The Golden Bough,* are also of interest here. Frazer was not a fan of what he called "primitive magic . . . the bastard art," but he did grasp the concepts behind it.

If we analyze the principles of thought on which magic is based, they will probably be found to resolve themselves into two: first, that like produces like . . . and, second, that things which have once been in contact with each other continue to act on each other at a distance after the physical contact has been severed.

Both branches of magic . . . may conveniently be comprehended under the general name of Sympathetic Magic, since both assume that things act on each other at a distance through a secret sympathy, the impulse being transmitted from one to the other by means of what we may conceive as a kind of invisible ether, not unlike that which is postulated by modern science for a precisely similar purpose, namely, to explain how things can physically affect each other through a space which appears to be empty.[5]

The ideas behind "sympathetic magic," which Frazer observed in numerous cultures including the folk traditions of our own, are, first, that if a plant looks like the thing it is used to treat, it will be effective in doing so. Jergon sacha is only one example of this. There is another Amazonian plant that is believed to be effective against abduction by mermaids, for instance, who take the form of pink dolphins and can drag the unwary to the land of the dead beneath the waves. The fact that this plant guards against such abductions is evident from its appearance—it is pink and grows in a straight line, each clump separated by a gap of some feet: like a dolphin leaping from a river.

In Haiti this principle is also known and used; for example, in magical works to "bind" a person (i.e., keep them tied to someone or hold them against their will), vines are commonly used to wrap a symbolic representation of the person to an object such as a stone. Vines are used as opposed to string, for example (which would do the job just as well), because a vine will naturally wrap itself around something else (such as a tree), and its strength of attachment grows as the years go by.

Frazer's observation that there is a "secret sympathy" to nature also suggests another shamanic principle: that each plant is not just representative of itself and its own spiritual qualities, but of its whole species. Thus, a lavender bush is not just one bush but all bushes, and not just one plant, but a gateway to the whole natural world. A shaman who makes an ally of lavender, therefore, is in contact not just with that plant, but with the soul of nature itself, and with every plant there is or has ever been.

The first step for every shaman, then, is to make contact with his or her first plant ally, which becomes a *punka* (a Quecha word that means doorway) for the world of plant spirits, as the shamans of the Amazon say. The exercises that follow will enable you to make contact with an ally of your own.

A Journey to the Plants

Whatever else the medicine man may specialize in (healing rituals or divination, for example), the key approach of all shamans, ancient and modern and across all cultures, is a special state of trance consciousness that gives him or her communion with the energy (or spirit) of the universe. This has come to be known as shamanic journeying.

To the Sora of India, journeying is understood as the shaman commanding his soul to leave his body so he can meet with the spirits and they can speak through him. In Siberia, the shaman takes flight to the otherworld to meet spirit teachers and rescue lost souls. In Haiti, the soul of the shaman goes to Gine (pronounced Ginn-NAY; primal Africa) so that ancestors and angel-like beings can take over his body and transmit their healing through him. In all cultures, then, journeying is a means for exploring the spiritual universe, making contact with tutelary spirits, recovering lost energy, or finding out more about ourselves and our purpose. In this exercise we use it to meet the spirit of the plant that is calling you.

To take any shamanic journey, find a time and a place where you can be alone and undisturbed for twenty minutes or so, then dim the lights or cover your eyes and lie down on your back, legs outstretched, arms at your side, and make yourself comfortable. This is the classic shamanic posture for journeying.

Most journeys are taken to the sound of drumming, which encourages dreaming patterns to emerge in the brain, taking the shaman deeper into a more holistic experience of the world in its fullness. You can drum for yourself (although you will not be able to maintain the posture, of course), have a friend drum for you, or use a drumming tape or CD to guide your journey. All are effective.

Expressing your *intention*—the purpose for your journey—and keeping this in focus is also important. Intention is the energy that guides the journey and ensures that you do not wander aimlessly. So the next thing is to express your intention by framing a positive statement of purpose. This opens up your inner awareness to engage with the mind of the universe so it can work with you and begins to direct your energy toward a specific goal. It also means you don't get distracted because you have a clear and definite objective in mind.

On this journey, your intention is *to meet with a plant ally*—the consciousness of a plant that will guide you into the magical world of the collective plant mind. You do not need to have a specific plant in mind. Stay open, instead, to whatever comes to you.

As soon as the drumming begins, imagine yourself entering a place that connects you to the Earth in a way that is meaningful to you—perhaps a desert, a jungle, or a mountaintop. All of these are fine as long as you feel immersed in this place and as if you are really there. Then express your intention again and allow your imagination to take you where it will. Don't try to control or dictate what happens or the information you receive. Just relax and be guided by intention. All you need to do is receive.

When your ally appears to you, spend some time in conversation with him or her. Inquire about the plant's healing gifts and the way these properties manifest in the plant itself. Ask how you can work with your ally and the plants that embody him or her. Also ask if there is anything you can offer in return for this help.

Shamans say we must feed our allies to give them the energy they need to help us. This "feeding" may take many forms according to what that spirit needs—a ritual, for example, or an offering of some kind. Be guided by the plant as to what it requires.

If you are new to journeying, this may all feel a little strange to you. Many people, after their first journey, ask the same question: "Was this real (i.e., did I really meet a plant spirit) or was it just my imagination?"

There is no *just* about it. Even if it *was* your imagination (and the imagination is never to be diminished or equated with "fantasy"), you will have learned something important from deep within yourself, from a place of untapped potential and natural power that we normally do not use in daily life. This place of power is also the world of spirit.

So you can conceive of this journey however you like—your imaginative genius at play or a communion with the universal mind—because it is the outcome that is most important: if you receive useful information from your journey, then of course it is real. One way or another, the information you now have validates your experience and *makes* it real. In fact, by "tuning in" to the plants in this way it usually becomes obvious that we know far more about their qualities than we think; we see that plant medicine is not the preserve of doctors, herbalists, or other "experts," but part of the ancient cellular wisdom of humankind and the birthright of us all.

When you have spoken with your plant ally, ask if you may merge with him or her (the answer will almost certainly be yes), then reach out your arms and feel the energy of this spirit enter your heart and become part of you. Breathe it in and allow it to permeate your whole being.

When you are ready, or when the callback sounds on your drumming tape, retrace the route you have taken to this otherworld meeting place and come back to ordinary consciousness.

Meeting the Plant in this World

Once you have met your ally in spirit-space, you will know which plant to work with in ordinary reality. Quietly now, in a meditative and focused way (what shamans call a walk of attention), go out into nature and find that plant. You won't need to search for it; just enjoy your walk and listen for its call. *It* will find *you*.

Spend a little time with the plant once you have met it, merging your consciousness with it and sensing its qualities and healing intentions. Allow your eyes to go a little out of focus as you look at it so you can see its essence beyond the distractions of its form. This is a shamanic technique known as *gazing*.

After a little while of this, begin to write, as a stream of consciousness, all the ideas and associations that come to you as you look at or think about this plant. This is another way of knowing it—by allowing it to speak through your intuition.

Knowing Your Ally

Gather together all the information you have received—from your journey, your gazing and meditation on the plant, and your stream of consciousness writing, as well as any additional research you have chosen to conduct on this plant. Then begin to refine it so you come up with a "specification" for the ally you will be working with, similar to the example in the box below. (The section on spiritual intention is the most important, as this is the information the plant itself has provided during your journeys and quiet reflection on its qualities.)

An Example: Sage

Collection: Leaves can be harvested at any time but are more
potent in spring.
Disinfects open wounds, and can also be used in a burner to
cleanse the air, so was once used in hospitals to kill viruses.

Sage is cleansing to the mouth as well and contains essential ingredients used in mouthwashes and toothpastes. It will heal throat infections, ulcers, and tonsillitis.

Sage calms diarrhea and reduces sweating and body heat (e.g., hot flashes during menopause).* It dries up the milk of a breast-feeding mother. (To wean baby off the breast the mother should take sage. Its properties will be transferred to the milk, but this is quite safe for baby.) It removes mucus congestion and cures or prevents colds.

Spiritual Intentions—What the Sage Has to Say: "I am the quality of *drying out*. Most of my healing has to do with this. I help when there is too much fluidity—which may also mean change, confusion, emotional upheaval, or uncertainty. I am grounding at such times and offer cleansing and protection. Drink me, breathe in my smoke, or hang me above the doorway of your home, and I will purify you and prevent unhelpful energies from coming in. Your dreams will also be more powerful when you sleep with me and will reveal your future. Some say that my fragrance is the aroma of wealth, and I will draw money to you. The shape of my leaves is the shape of your tongue, so use me for infections of mouth and throat. My name *(sage)* means wisdom, and I will bring you knowledge and strong memory."

Cautions with Sage: Do not use internally during pregnancy. Avoid with epileptics.

---

*NOTE: Do not, however, use sage to reduce sweating in fevers.

Once you have assembled your data on the plant, check your intuitive sense against the information in an herbal encyclopedia to see how your assessment compares with that of others. You may be surprised at how accurate you are. But, in fact, this is not so amazing after all, if you think about it—sometime, somewhere, the spirits of the plants told someone their secrets so they could be recorded in an encyclopedia at all—so why shouldn't they speak to you?

SHAMANIC PLANT HEALING

The comparatively limited concept of a plant "medicine"
as a substance that possesses or is reputed to possess
curative or remedial properties is restricted to Western
society. In many areas of the world the definition is
extended to include medicine that makes people fall in
love with the user, medicine for winning sales, medicine to
appease the ancestors, or anything that is unusual, sacred
and imbued with magical power.

FRANK J. LIPP, ETHNOBOTANIST

Shamanic healing with plants is hardly ever—and certainly never solely—about administering medicine in a form that a Western doctor might understand the term. Instead, it may include divination, the receipt of spirit blessings, magical potions to change "luck," or the healing of the soul through the energy of the plants, and not through their physical attributes at all.

The aim of a plant shaman, in fact, is often not even to cure a condition at all, but to remove its spiritual cause by restoring in the patient a sense of balance, harmony, and reconnection to the sacred and the Earth. This balance and at-one-ment is regarded as the natural state of human beings; once the patient experiences it again, the illness (which was a messenger of disconnection rather than a condition in itself) has no need to remain and will magically disappear. The plant is an intermediary in this, playing the role of doctor, counselor, confessor, therapist, or friend—whatever the patient or the shaman needs it to be, in fact, in order for balance to be restored—rather than a source of medicinal or chemical properties.

As ethnobotanist Frank J. Lipp, PhD, points out, plants "play an integral role in ideas of balance and cosmological order that often reflect sophisticated medical theories of the human body, the symptoms it experiences and their underlying causes. There is, moreover, no hard and fast distinction between 'medicinal plants' and 'food plants', since many plants, such as maize, chili peppers and sage, are utilized both as food and medicine."[6]

It is the shaman's intention and the cooperation of the plant that decides whether it is a healer or a foodstuff. In Andean Peru, for example, coca leaves are used medicinally to give people energy and to help them cope with the high altitudes that can drain the body of strength, but they are also among the most sacred plants used as offerings to the gods. It is not unusual, therefore, to see an Andean shaman chewing coca as a medicinal food while at the same time offering the leaves as part of a ceremony. At one and the same time, the plant is a medicine, a food, and a spirit ally. What makes the distinction is the shaman's intent.

Another example, this one offered by Lipp, is that of the healers of Zaire, who must visit the sacred woods where the ancestors are buried and pick certain herbs just as the sun's rays fade, if they wish to cure a patient of fever. "A plant's medicinal potency may lie dormant until the requisite incantation has been pronounced which will define its purpose and direct its action," says Lipp. "The ancestral spirits are petitioned to make the fever's heat fade in the same way as the light of the sun." Without this petition—this statement of intent—the herbs would not work at all, no matter how chemically potent they might be.

PLANTS AS PLACEBOS

As Lipp also remarks: "Since the time of Plato it has been recognized that substances with no inherent chemical efficacy are nevertheless useful in eliminating symptoms and pain. . . . It has been estimated that 35–45 percent of all prescriptions* are for drugs that by themselves could not affect the conditions for which they are prescribed."[7] We know this as the placebo effect—the use of an inert plant material such as sugar, which is given to a patient with the promise of a successful cure—and are rather dismissive of it in the West, but the fact is that *it works*—better than orthodox medicine in many cases.

From Lipp's estimate, our recovery from illness relies on this effect in a third to almost half of all cases. If there is no medicinal quality to the drugs we are taking, what else can it be that accounts for the ability of these plant derivatives to heal us, but something nonchemical—*the spirit of the plant*—and the intention of the healer to heal?

Perhaps the biggest problem for the Western mind in accepting plant spirit medicine as a bona fide practice is, as we said earlier, that its outcomes are not scientifically measurable. It is not the same as giving someone an aspirin and recording that thirty minutes later the problem is gone—that the *symptom*, at least, if not the real problem (the underlying cause of the headache) is cured.

How do you measure, for example, the effectiveness of a plant in making someone fall in love with you or increasing your business turnover? Because of the model science uses, there will always be other factors to look for so that "supernatural" causes can be ruled out. That, after all, is its goal. Science is tautological in this respect, because whatever we look for we will, of course, find. Whenever scientists set out to find an alternative explanation proving that plant spirits do not exist, that is what they discover since they were, all along, seeking this proof. The irony is that even when a placebo is shown—scientifically—to work, we tend to deny ourselves its healing by calling it ineffective in its own right and putting its success down to gullibility. If we open our minds to how plants really work, though, what we notice is the possibilities rather than the limitations, the miracles, not the "trickery" involved.

---

*These figures refer to prescriptions in the West.

PLANTS AS A MEANS TO BALANCE

In her article in *The Globe and Mail* newspaper, "Bewitched in Bolivia,"[8] Michele Peterson provides insight into one form of shamanic healing, which, again, does not rely at all on the ingestion of plants as a medicine. Here, instead, the plant has a counseling and therapeutic use and heals through the restoration of balance. Peterson writes:

> A full moon was rising over Mount Illimani as I stepped into an alley in the Bolivian capital. I was a few paces behind Juan Mamani, a mountain magician, and we were looking for a deserted corner where he could perform a psychic diagnosis through the casting and reading of coca leaves. . . .
>
> Finally, in the dim shelter of an adobe wall, he unfolded his stool and lay his *botiquin*, or medicine pack, across his knees. Emptying a small plastic bag of pale green coca leaves, he packed several in his cheek and began to chew. Then, he placed one flat leaf on top of a weathered coin, made the sign of the cross, held his hands to his abdomen, breathed deeply and exhaled lightly onto the leaves. Lifting his hands high, he let the leaves scatter. My reading had begun.

Michele had been alarmed by frightening dreams about her family and had sought out a mountain magician (known locally as a *kallawaya*) to bring her peace of mind through an understanding of these dreams. At the start of the divination, the kallawaya ripped a pattern into a coca leaf, telling Michele that it represented her father. He then cast this to the wind and noted where it landed. "He is healthy. But he is disconnected from Mother Earth and lives in fear," he said.

Another leaf represented her mother. This one caught the breeze and landed upside down. "That means sorrow," he said.

Then, looking at both leaves, their relationship to each other, where they had landed, and noting that they were in the center of a number of other leaves, the kallawaya concluded of Michele's parents: "They are surrounded by negative energy. They must reconnect with Mother Earth."

Coca leaves used in this way, while referred to as divination, are, in fact, diagnostic and healing tools, because once the cause of the problem is known, a cure can be found. In this case, Michele's concerns about her family would be alleviated and her parents helped if they could reconnect to the natural world.

The kallawaya were once doctors to Incan kings and are still consulted by 80 percent of Bolivians. Conventional medicine has also begun to recognize their knowledge. Kallawaya were the first to use the bark of the *cinchona* tree to control malaria, for example, and this is now the source of quinine, still one of our most effective ways of preventing and curing the disease. According to Michele, other "recent studies report signs of medicinal potency in several kallawaya medicinal herbs."

There the similarity with modern medical practice ends, though, because "the kallawaya believe that physical illnesses originate from the soul and are caused by the *ajaya*, or life force, leaving the body. The healer's job is to coax it back into the body and restore equilibrium with the spirit, as well as the environment." Drugs are not the answer; the goal is the restoration of balance through the action of the plant on the soul. (Also see chapter 4 for more information on this form of soul healing.)

In the Bolivian as in the Peruvian Andes, one way of finding this balance is by making an offering—consisting of herbs, money, and talismans—during a ceremony to Pachamama, the spirit of the Earth. Known locally as an *offerenda,* the ceremony is often made at sacred sites, called *huaca.* These are hills, lakes, and rocks that have special powers associated with them and are the homes to particular *apus* or spirits.

Michele's own service required a yellow candle, copal incense, limestone effigies, *madera del santo,* colored alpaca string, and several herbs, which she was told to burn and then to bury their ashes in the earth.

Being a Western journalist, she didn't do this at all, of course. She got her story and left. But if she had done as the kallawaya suggested, she might have noticed some unusual and beneficial results.

On a trip we facilitated to Peru in 1998, for example, we arranged for a group of Westerners to take part in an offerenda very much the same as Michele's. As we gathered on the mountainside near one of the area's sacred sites, a light wind began to blow, lifting the smoke from our small fire high into the sky. One by one the group members stepped forward to whisper their prayers to the smoke and to place within their offerenda package one or two coca leaves handed to them by our shaman, Doris Rivera Lenz. Then they burned their offerings in the flames.

This simple ceremony was life changing for some. Mysterious health problems that had plagued them for years suddenly cleared up, with no other form of treatment. Others found the strength to leave jobs that had been unfulfilling but which they had stayed in through fear of change. One couple who had only recently met decided, then and there, to get married. They did so on their return to the UK, and at the time of this writing (seven years later) are still happily together.

"I'd say it was the most important moment of my life," said one woman who gave up her teaching job as a result of this ceremony and later returned to Peru to study plant medicine with the shamans. She writes, "The divination revealed what I should do with my life—which I *knew* deep within me to be right—and when I made my offering I saw it as a real commitment to be true to myself and my calling. When I returned to England, nothing seemed the same and I knew I had to get back here as soon as possible. I've been living in Peru for some time now and I'm the happiest I've ever been."

If all of this can be accounted for by the placebo effect, then bring it on! We'd like to see a Western medical doctor do better in so short a time!

DIVINING WITH COCA:
AN INTERVIEW WITH AN ANDEAN CURANDERA

The Incas regarded coca as *the* divine plant, mainly because of its ability to impart endurance, and its use was entwined with every aspect of the life, the art, the mythology, and the economy of the Incan Empire.

Millions have chewed coca on a daily basis and the practice has remained for hundreds of years. It continues as a custom, not because coca (the basis for cocaine) is a "habit drug," but because it is central to Andean culture. Even today, the people measure distances in *cocadas*— how far a load can be carried under the stimulus of one chew of coca.

Andeans chew coca just as they do everything else: ritually, deliberately, and systematically. They carefully choose a mouthful of leaves from an exquisitely woven *chuspa* or coca bag and chew *lliptia** with the leaves to liberate their active ingredients.

But the ceremony that really brings out the spirit in the leaves is coca divination. Doris Rivera Lenz is an Andean curandera (shaman) who is expert in its practice. In the following interview, she offers insights into the nature of healing and illness, and the role of plant spirit medicine in this.

What is coca divination?

DRL: *Divination is meeting with the spirit of the element that you are working with, whether it is coca, maize, or a mountain. In the case of coca, you meet the mother spirit, soul, or power of the plant, which is the sacred part that never dies.[†] The practitioner must be in total communication: spirit-to-spirit. It is more like listening to the coca leaves than reading them. It is a higher state of consciousness and you have to be prepared to integrate yourself spiritually to help another spirit.*

Human beings are sacred cosmic seeds in evolution. The coca is a sacred seed like us, only of the vegetable kingdom. It has been created by the Earth to guide and heal its younger brothers: ourselves. Similarly, we have been created to help other people. As we become more open, we discover plants like coca. Not everybody sees the spirit of coca, but it is here to help us.

---

*Lliptia is a strong-tasting "gum" made by burning the roots of the quinoa plant to produce a substance that is extremely alkaline. When chewed with coca it helps to break down the leaves and activate their alkaloids. A similar substance, called *kale* (or "lime" in Spanish), is produced by heating limestone.

†In Peru, the *mother spirit* of a plant is the soul of its species. Each plant, each leaf of coca, for example, contains its own soul, but it is the combined energy or essence of these leaves that forms the mother.

What is the cause of disease, and how is it cured by the spirit of the plants?

DRL: *Illnesses do not exist. We create them with our minds, according to our attitudes and the things we do. Resentment, for example, causes cancer. A woman whose ovaries are unwell [with cancer] may be resentful and [so] suffers trauma. People who do not have the freedom to express their feelings suffer from throat problems, and so on.*

So how do we heal them? First we need to look at them through the coca leaves, to know what has happened. Why are they resentful, fearful, or anxious? What is causing their problems? Difficulties existing outside their bodies, such as a theft, disillusionment, or being lied to, may affect them because they are predisposed to have this pain. Such people get ill because they are not in equilibrium with themselves. The coca shows when and how this began; it tells the story of how they got ill.

Human beings are always predisposed by their attitudes. This is why you need to know their story. Someone who has a superiority complex or is aggressive and violent is on a downward spiral. They are weakened in their heart, stomach, and solar plexus: the ñawi or naira, where emotional attitudes are held.* In the Andes, people will frequently consider an aching stomach to have been caused by sorrow.

A person who harbors feeling of hate may feel perfectly well for a time but problems with their children, their husband, or lack of money, intensify their emotions, which degenerates their body on a cellular level. So they create their illness because they are already out of equilibrium.

Can you explain the concept of the ñawi and how it relates to illness?

DRL: *In Quechua it is ñawi, or in Aymara, naira. It means eye, or energy center of the body, but chacra is also a very common word in Peru, and is Quechua for a piece of cultivated land or field. I believe it has the same linguistic root as the Hindu chakra. Just as some fields have lots of stones, and others are very fertile, so our bodies, also part of nature, are similar.*

---

*The *ñawi* or *naira* are the Andean equivalent of chakras.

Less than a generation ago, people would make offerings before preparing their fields for sowing. They would chew coca leaves, drink chicha or maize beer, and play music—a whole ceremony. The ancient healers or shamans would give floral or smoke baths to people, curing them of illnesses, fright, and so on—the health of the land and the people were treated as interrelated. People identified themselves with their fields and with nature. So when I remove negative emotions from a person, it is like I am removing weeds from their chacra/field.

When they are feeling desperate, the people of the Andes benefit from going to a wild place or some ruins, to scream and shout so that even the mountains will hear! They align with natural forces; this puts them back into equilibrium.

So, do people come to you for coca divination because they are unwell? Is it more than "divination" as we would understand it in the West?

DRL: The majority are unwell in their spirit or mind; there are lots of problems today. They are particularly afflicted in the stomach, the place of emotional pain, and also where we are joined to life.

The first thing is to discover what is going on: the wife had an accident, the husband was unfaithful, they haven't got a job, the house is falling down. . . . Then I look to see their capacity to accept criticism, to listen to the mother leaf ticking them off saying: "You have done this, you are insecure, weak, a drunk, or a prostitute." What is the story? Is it karmic—or something they are doing?

That sounds like a psychological approach—what people are doing to themselves. How do you make sense of the belief that some problems are caused by sorcery?

DRL: I show the person that he is not the victim of sorcery and is creating the problem in his mind. Talking about it brings it out and is the first part of becoming well again. It is true that some people will take vengeance through black magic when they feel prejudiced or offended in some way,

because they are sick. When people think they have power and feel supe-
rior, the ego can become very negative.

The first thing I do is to wake up the consciousness of the person who
has been harmed and tell them that evil does not exist! "You are inventing
it," I tell them. I need to use a bit of psychology!

Black magic does not exist then?

DRL: Neither good nor bad exists; it is a universe, and we create the good
and the bad. But I recognize that the person may feel attacked, so when
someone falls ill it means they are weak and I as their curandera* must
speak positively and encourage them to shine light on it. Then they can
create positive thoughts for themselves. If I agree with them and say they
are bewitched it just makes them worse.

But do you believe that such magic can exist?

DRL: Of course! But the act itself is not so powerful as white magic. It is the
negative attitude of the black brujo† which creates the power of the spell.
If he gets hold of a chicken and takes off its feathers, puts a toad inside,
and hangs it in the doorway of a hated neighbor,‡ he can give that person a
nasty fright, but without a powerful negative attitude nothing will happen.
But if his intentions are very negative and the person he attacks is weak,
then that person will become ill quickly.

The most powerful brujos are found in the jungle where there are
plants for healing, just as there are dangerous plants that can paralyze
your body and so on (see box on page 32). But plants have much more
wisdom than people. Do you think that if I go to a floripondio§ and say "I
want help to do harm to so and so," that their plants will automatically be
at my disposal? No! You have to make a pact with their spirits.

---

*A *curandera* is an Andean healer. Curandero = male; curandera = female.
†A *brujo* is a sorcerer who harms through negative energy, often on behalf of a client.
‡An Andean form of cursing.
§A *floripondio* is a shaman who works with flowers.

What Can Heal Can Harm

It is a recognized fact in most shamanic healing—as in modern science—that plants that heal can also harm. In the Amazon, for example, there is a tree—colloquially known as the 'brujo (sorcerer) tree'—that can be used for positive magic. If a photograph is nailed to it of someone to be harmed, however, through the intentions of the sorcerer, the tree becomes a malevolent force and its vicious thorns, as well as the scorpions, snakes, and poisonous spiders that live in its branches, will ensure that the victim suffers.

This knowledge that "what can heal can harm" is not confined to the Amazon. The folklorist Zora Neale Hurston, in her book, *Tell My Horse: Voodoo and Life in Haiti and Jamaica* (New York: Harper and Row, 1990) writes that the Jamaican God Wood tree (birch gum)—"the first tree that ever was made . . . the original tree of good and evil"—also has these powers. A *bokor* (sorcerer) wishing to do harm will drive a rusty nail into the tree trunk while thinking of the person to be injured, and that person will grow weak and die. A small piece of bark from the God Wood tree applied to the brow of a person who is sweating, it will also kill him. The physical toxicity of the tree, through spiritual intention, is therefore transferred to the victim.

All plants have the power of both life and death in this way, depending on the intentions and skill of the shaman: "Boil five leaves of *Horse Bath* and drink it with a pinch of salt and your kidneys are cleaned out magnificently," says Hurston. "Boil six leaves and drink it and you will die. *Marjo Bitter* is a vine that grows on rocks. Take a length from your elbow to your wrist and make a tea and it is a most excellent medicine. Boil a length to the palm of your hand and you are violently poisoned."

If "magic" is mainly in the mind, do people also need to believe your healing ceremony has done something in order for it to work?

DRL: *When people* trust *that I am a white curandera they open up. Then I have special permission to go into their soul and work with suggestion.*

Let's say I give them a bath in an herb with spines and ask permission from the spirit of that plant to heal the person with fright or a bad spell—I bathe them, I put them on a diet, I cleanse them and purify them. I call their soul and give them strength and they will get well. But, yes, I also talk a lot! There are times, though, when I don't say anything.*

What is different about people from the West? What do they need?

DRL: Their heads cutting off! No, it's only a joke!

Their spirits have weakened; their religion has failed them and the church authorities have kept vested interests and institutions going. Eventually people have thrown the baby out with the bath water.† The truth is that we are gods and we should believe in ourselves first.

As we know, all gods come through nature. But what has become of Western religion? Materialism, loss of identity, loss of customs. There is so much struggle today. People are no longer thinking about nature, but about money and the help they need. They have become completely insecure. Imagine if we went to live in nature again, surrounded by mountains, or in the rainforest, how much more healing it would be. Yet the tendency today is for everybody to want to move into cities, to live like Americans, build motorways. It's sad.

I've spent time with people in the Andes. I have seen people leaving their traditional clothes and customs. They say "Why do you believe in the Earth, the Sun, the puma and the condor?" They go to the city and see a TV and think, "What a beautiful TV!" They sell their llamas and buy one. I am sad to see their children, who are so pure, being contaminated in this way.

They learn negative habits, too, and are hypnotized, and no longer want to work their land. It really hurts my soul to see them obsessing about dollars and forgetting their power. This loss of values [and the increasing desire for material things] is happening so fast, it's incredible! But it's the Western influence which has been working over five hundred years.

---

*That is, there is nothing to say; the plants dictate what to do, or do the work for the healer.

†Meaning, people have rejected religion and lost touch with spirituality in the process.

People will get a nasty shock from seeing the increasing changes and natural disasters on the Earth and we will be shocked into changing.

And is that the only way we can change—through shock and fear?

DRL: *Unfortunately, yes. We need to work daily to balance ourselves, so the collective fear will not infect us. Even if those around you are overcome, you must maintain your center.*

Everybody worries about their future, no? But there will come a time when no one will want to consult about it any more, they will have finally woken up to the realization that there is no future in the way we are going. They will be shocked into living in the present and this will create a new human being. We will realize that individualism doesn't work and this will unite us in a shared future.

Desperation will show the necessity of love. Who will want to do harm when money and material things have become useless? We will come back to a new kind of community consciousness. We are beginning to anticipate this and becoming more conscious, but we are swerving about. There is so much wisdom in nature, she rears us like her children, teaches us to ask permission, to care for her like ourselves. Nature is where we should start.

Interestingly, Doris, an Andean curandera, made these last remarks in 1998. In 2006, the celebrated environmental scientist, James Lovelock, said exactly the same thing, suggesting—happily, and at long last—a convergence of opinion between science and shamanism, even though that opinion is in respect of a rather unhappy state of affairs. In an article in a national U.K. newspaper, Lovelock writes that, as a result of our materialistic values, climate change is now so far advanced that it will never be reversed. Instead, we must come together to find a new way of life in the knowledge that the world as we know it has no future. "Let us be brave and cease thinking of human needs and rights alone, and see that we have harmed the living Earth and need to make our peace with Gaia," he writes. "We must be strong enough to negotiate, and not a broken rabble led by brutal war lords."[9]

INFORMATION FROM THE OTHERWORLD

Divination as a holistic healing system is not restricted to Peru, but is practiced in every traditional culture that recognizes the sanctity of nature and its ability to teach and restore balance. The Ifá system, which originates with the Yoruba of Nigeria, for example, shares many of the same methods used by the Peruvian shamans. Ifá, too, uses plants as a doorway to the gods.

Ifá divination takes three main forms: *Obì,* which uses kola nuts; *Òkpèlè,* which uses a divining chain that has eight half-shells of the òkpèlè seed strung from it; and *Ikin,* which uses palm nuts. The diviner of the Ifá oracle is called the *babalàwo,* also known as the "Father of Secrets."

Sixteen nuts are thrown, and the babalàwo records the outcome of the throws on an *opón,* or divining tray, using *ìyeròsùn* (termite dust from the *ìròsùn* tree). Eventually an *odù,* or mystical story, is traced. There are 256 different odù, each with proverbs and verses to them, and it is these the diviner interprets in line with his client's question or healing need.

Although Ifá and the interpretation of odù are skilled arts, those who use this system within the Santeria tradition* believe that the diviner's connection with the plants (the "mouths of the gods") are what is most important. "No one knows all of these odù or what the stories mean—and new interpretations are made every day as each story is added to," said one priest of Santeria. "But as long as the diviner is in good standing with the gods and at one with the plants, he will find the answer his client needs."

The outcome of an Ifá divination can often seem curious or confusing, as plant spirits tend to speak in parables and poetry rather than direct instruction (as you may have found). Nonetheless, powerful healing and

---

*Santeria is a form of Vodou arising from African slave beliefs, mixed with the beliefs of the Catholic slaveholders and the Cardec spiritualist movement that became popular in America in the 1800s. It is most associated with Cuba and, nowadays, with areas of mainland America, such as Florida, that have a sizeable Hispanic population.

protection will result if one follows the advice of the spirits—no matter how strange it seems.

There is a story (probably apocryphal), for example, of a man who visited a babalàwo for a reading because he wanted protection on a journey he was about to take. Instead of straightforward advice, the spirits told him to cast a number of coconuts and a length of rope into the ocean. Even though this sounded like nonsense and bore no resemblance to his request, the man did as he was told and the next day set off on his journey.

A few miles into his walk, he was set upon by thieves who stole his money and threw him from a cliff top into the sea. The man was afraid because he could not swim; but as soon as he surfaced, his hand came to rest on the bundle of buoyant coconuts he had thrown into the ocean just the day before, around which the rope had become entwined to hold them all together. By clinging to the buoy, the man was saved from drowning and drifted safely to the shore.

In this case, the healing from the plants had nothing to do with "medicine" as we understand it, but was nonetheless powerful in saving the man's life and changing his luck.*

A Celtic Tale

The Ifá system, as we see, uses nuts. Taking a cross-cultural perspective, we learn that nuts are containers of wisdom in the tales of many different traditions. In Celtic myth, for example, the Salmon of Knowledge, Fintan, swam in the Pool of Wisdom in a grove of sacred hazel trees.

The Druid Finegas, who had guarded the salmon for seven years, decided one day to eat it, aware that whoever did so would gain all knowledge. As the salmon roasted, Finegas left his pupil Demme in charge, but in turning the fish, some of its oil spat onto Demme's thumb

---

*For more information on Ifá divination, see Ross Heaven, *Spirit in the City* (Transworld/ Bantam Books, 2002). See also Ócha'ni Lele's books on Afro-Cuban divination, *The Diloggún, The Secrets of Afro-Cuban Divination,* and *Obi: Oracle of Cuban Santería,* all published by Destiny Books, Rochester, Vermont.

and burned it. Without thinking, he put it in his mouth to soothe the burn and, in that moment, received all wisdom.

Finegas recognized the change in his pupil as soon as he returned and advised him to join the warrior heroes of the Fianna. Demme did so—becoming the wisest and greatest hero of them all: Fionn Mac Cumhal, father of Oisin. It was the hazelnuts, of course, that made the salmon wise, and when Demme sucked his thumb, it was their power that brought him knowledge.

HOODOO PLANT MAGIC

The Hoodoo tradition of the American South is another that works with plant energy while often not requiring a patient to "take" a medicine to experience healing.

The word *Hoodoo* has African origins, and is used to describe various forms of magic, folk healing, and hexing, using roots and herbs. Nowadays, the Hoodoo practitioner is often referred to as *root doctor* or *juju man.* These rather quaint terms belie the power of the tradition, however, because in Africa, these practitioners were priests *(botonons)* and sorcerers *(azondoto)* who were rightly feared and respected for their herbal expertise and knowledge of the spirits and *bochio,* or soul.

When these priests arrived in America as slaves, they brought their knowledge with them, but the unavailability of some plants meant they had to adopt Native American and even European plant allies and practices in their work. Within modern Hoodoo, we therefore see the fusion of many plant magic strands.

One of the best-known forms of this magic is the *mojo bag*—a bundle of plants and consecrated items made to bring luck and protection or to ward off negativity. These are not so different from the offerenda packages created by Andean shamans as a call to spirits to direct their healing energies toward the person making the offering. (See a photo of Doris Rivera Lenz at a ceremonial Andean offerenda on page 3 of the color insert.) The word *mojo* comes from the West African *mojuba* (prayer) and denotes a method of directing spiritual energies to similar effect.*

To look at, a mojo is a flannel bag containing magical items, which is usually carried on the person, tucked away out of sight and often worn next to the skin. If the mojo is intended to protect a property rather than a person, it may be hidden near the front door. But in either case, it is important that no one sees or touches the bag apart from the person who owns it, or its magic may be lost.

The contents of the bag vary according to its purpose, but typically there are at least three magical objects. These might be roots, leaves, feathers, crystals, stones, snake vertebrae, lodestones, metal charms, or papers on which *sigils* (magical symbols) have been drawn or wishes written. Sometimes there are more than three, in combinations that always add up to an odd number.

The reason odd numbers are used differs according to whom you ask, but a popular idea is that the universe operates on odd numbers in order to keep things moving. Three, for example, wants to "tumble into" four, whereas four is "solid" and fixed rather than flowing. To get energy moving to the benefit of a client, therefore, or to undo a run of bad luck that seems never ending, the magician uses an odd number to signal to the universe that now it is time for a change. The Jungian psychoanalyst Marie-Louise Von Franz also discusses this in her book, *The Interpretation of Fairy Tales.* Quoting Jung's paper, "A Psychological Approach to the Dogma of the Trinity," she remarks that "three is generally connected with the flow of movement . . . for movement you need two poles and the exchange of energy between them—for instance, the positive and negative electric pole and the current which equalizes the tension."[10]

An example of an old mojo trick for financial success is to wrap High John the Conqueror root in a dollar bill and add frankincense and a little sugar, then tie them all in a green bag. High John the Conqueror

---

*Mojo bags are also known by other names, including "lucky hand," "trick" (or "root") bag, and *gris-gris*—from the African word *gree-gree*, meaning "charm" or "fetish."

is a mainstay of African-American magic and is also part of legend. It is named after a proud African slave, High John, who refused to be servile to his masters and, for this reason, is associated with the ability to overcome problems and stand up to oppression. The root is also used in sex magic because, when dried, it resembles a man's testicles. High John is usually quite easy to obtain in the West, but if it does prove difficult, a nutmeg can be used instead.

Notice that here, five elements are used: the root, the dollar bill, the frankincense, the sugar, and the bag itself. The symbolism behind these items works in accordance with the plant spirit principle that "like attracts like," so that High John is used to "conquer" money problems, and the higher the denomination of note used, the more money will be drawn to the magician or his client. The bag is green because this is the color of money (greenbacks), and sugar "sweetens the pot."

For a peaceful home, angelica, olive leaves, rosebuds, lavender flowers, lemon balm, and basil leaves are tied into a blue bag (for harmony) along with a few intertwined hairs from all family members. Again we see the doctrine of signatures at play in the choice of angelica ("angel root") and balm for spiritual harmony and the leaves of an olive branch for peace.

A more unusual mojo trick is for invisibility. This requires poppy seeds and dried fern leaves to be ground together "beneath a dark moon." To this are added myrrh, marjoram, slippery elm, and fresh dill, mixed with spring water and almond oil. When it is dry, the mixture can be added to a mojo bag or sprinkled over objects to conceal them from others.

The magic in a spell like this, as Doris Rivera Lenz says of her own work (see the interview earlier in this chapter), may be to give a client the *self-belief* that she will not attract attention to herself, so she can navigate her difficulties with more confidence, knowing that she is unseen. The change, in other words, is to the client and not the external world.

Once it is prepared, a mojo bag is purified in incense and its spirit "fed" with rum or whisky and with Florida Water. Body fluids may also be used, especially if the charm is to influence another person. To make someone fall in love with you, for example, it is useful to have a little of her sweat, urine, or saliva to dab onto the bag (you can also place other personal items—the closer to her DNA as possible, such as hair or nail clippings—into the bag itself).

A mojo bag can also be used for divining (as we understand it in the West) by attaching it to a string and using it like a pendulum. By asking the spirit of the charm to show you *yes* (often by spinning clockwise) and *no* (by rocking backward and forward), you have a spiritual device for answers to simple questions. For more complex questions (for example, "How do I make the man I desire fall in love with me?"), it is better to write out the letters of the alphabet and let the pendulum spell out the answer by gravitating to each letter in turn.

THE ESSENCE OF HEALING

How is it possible that plants can affect human beings, situations, circumstances, and life energies like this—remotely, as it were? That is, without being used as a form of curative for a specific medical problem, but more as a harmonizer, a magical attractant, or a conduit for spirit, energy, or luck?

Let's ask Cleve Backster, a scientist working in the unlikely field of lie detection and interrogation techniques, whose job was to teach policemen and security agents how to use polygraph equipment and interpret its results. Cleve Backster's work is recounted in detail in one of the classic books on the plant world, *The Secret Life of Plants*, by Peter Tompkins and Christopher Bird.[11]

Backster decided one day to attach the electrodes of a lie detector to the leaf of a dracaena plant to see if the device was sensitive enough to pick up reactions from a nonhuman subject. Probably not, he mused, but there might be some reaction if he burned the leaf to which the electrodes were attached. The second he thought this—and before he had even picked up a match—there was a dramatic peak in the tracing pattern on the polygraph chart, a trace signature that Backster would eventually come to recognize as fear.

Intrigued by this, Backster continued his research, testing almost thirty different plants in the same way: by attaching electrodes to them and then thinking of some action he might take toward the plant. The results were always the same. It was significant that the plants reacted *before* any action was taken, leading Backster to conclude that not only are plants as sensitive as human beings (or even more so), but they are also able to read emotions and intentions, because there is a form of psychic connection, or affinity, between plants and people.

As his work progressed, Backster realized that plants react not just to threats, but to presences or movements in their environment. He demonstrated to a group at Yale, for example, that the intended movement of a spider in the same room as a plant caused changes in the trace patterns of a polygraph to which that plant was attached. The plant had a precognitive sense of the impending and was attuned to intention before the movement itself. "The spider's decision . . . was being picked up by the plant," said Backster. "They [plants] seemed to be attuned to animal life."

Backster's other results show that plants have memory, emotions, and very humanlike reactions, as well as psychic abilities. In one of his experiments, six students randomly drew lots to see which of them would destroy one of two plants in a room. The person chosen would act in secret so that Backster and the other students would not know his identity. In fact, only the second plant would know who the student was because only it would witness the crime. After the plant had been destroyed, Backster attached a polygraph to the surviving plant and paraded his students one by one in front of it. The needle went off the scale when the student responsible appeared.

In a kinder experiment, Backster also demonstrated the love or empathy between a plant and its owner. One day he accidentally cut his finger and noticed that a plant being monitored was demonstrating a stress reaction of its own, as if it was experiencing Backster's pain and shock at the sight of his blood. Using this perceived affinity as the basis for his experiment, Backster walked to a different building some blocks away and directed loving thoughts toward the plant. The polygraph recording showed a heightened trace as the plant picked up his intentions.

To see how far such thoughts could be transferred, Backster asked a friend to send love to her plants while she was seven hundred miles away, and he recorded their reactions. By using synchronized watches and a prearranged schedule, Backster was able to prove that not only did the plants respond to their owner's thoughts at the moment she sent them, but they also felt her anxiety when her plane touched down at her destination. Even when the plants were locked in a lead container, the results were the same. Whatever created empathy between plant and human came from something outside the electromagnetic spectrum.

Another lucky accident led Backster to explore this further. One evening as he was about to feed a raw egg to his dog, he noticed that when he broke the shell one of his monitored plants reacted strongly. Curious to see what the plant might be reacting to and what feelings the egg might be transmitting, Backster attached another egg to a galvanometer, and monitored it for nine hours. What he got was a trace corresponding to the normal heartbeat of a chicken embryo—even though the egg was unfertilized. His conclusion was that there is a life force or energetic field that connects and is contained within all things.

Another researcher, Alfred Vogel, brought us closer to an understanding of this field when one of his students, Vivian Wiley, conducted an experiment of her own. She picked two leaves from a saxifrage plant and took them into her house. Each day she projected love toward one of them and the intention that it would live, despite giving it no water and simply leaving it on her bedside table. The other leaf she completely ignored.

After one month, Vogel went to her home to photograph the results. The leaf that was being ignored was dry and decaying, as you would expect from any leaf that had been out of water for that length of time; but the other was as fresh as the day it was picked, even though its circumstances were outwardly no better. Wiley continued her experiment for another month, and the leaf she directed her love toward remained alive all this time while the other one crumbled away.

The mysterious energy through which we communicate with plants is love and intention. These are the essence of the universe. "Man can and does communicate with plant life," said Vogel. "Plants . . . may be blind, deaf, and dumb in the human sense, but there is no doubt in my mind that they are extremely sensitive. . . . They radiate energy forces that are beneficial to man. One can feel those forces! They feed into one's own force field, which in turn feeds energy back to the plant."[12]

PRIMA DONNA PLANTS

Vogel also found that plants have unique personalities. Those with a large electrical resistance (the most powerful) are often difficult to work with (even "prima donna-ish"), whereas those with large leaves and high water content are more responsive to the mental power of intention.

This is quite in keeping with the findings of plant shamans. Ayahuasca shaman Javier Arevalo tells us in chapter 3 that ayahuasca—the vine of souls—is a "sensitive" plant, and to use it the shaman must follow a special diet where he denies himself certain foods and contact with others lest the plant should become "jealous" and work against him. A deep respect for the personality of the plant is essential to the visionary experience.

By the same token, medsen fey Loulou Prince tells of the precautions he must take when collecting orange leaves, another sensitive plant:

> When I find the leaves that I want, I go up to the tree with kleren*
> and talk to the leaves and pour some kleren on the ground. Each
> kind of leaf has a secret name, and I call the leaf by its name.
> Those things I never tell anyone. And there is a password to the
> leaf, and that is also a secret. But even if you don't know the
> secret name of the leaf, you respect it and call it by the name
> you know. For example you might say, "Orange leaves! Orange
> leaves! I need you—I take some but I don't take all. You come
> with me." And then you pick the leaves. [But] respect the plant,
> don't treat it roughly.

*Kleren is single-distilled cane sugar rum.

When you are done, you pay—you throw a few coins on the ground and say, "Here, I pay you!" and then you pour a little more kleren and say thank you. Then I wrap up the leaves and take them home and I light a candle and thank God and the spirits. Over the next few days, the leaves are spread in the sun to dry, if I am not going to use them fresh. And I take very good care of them and keep them clean, until I use them.

According to Vogel, this respect for the plant is essential for effective communication and empathy between our species. He remarked of scientists conducting experiments with plants that "If they approach the experimentation in a mechanistic way and don't enter into mutual communication with their plants and treat them as friends, they will fail."

It is essential to have an open mind that eliminates all preconceptions before beginning experiments. . . . Hundreds of laboratory workers around the world are going to be . . . frustrated and disappointed . . . until they appreciate that the empathy between plant and human is the key, and learn how to establish it. . . . The experimenters must become part of their experiments."[13]

THE VOICE OF THE PLANTS

A new form of lie detector—the Psychological Stress Evaluator (PSE)— was introduced to police work (and to plant experimentation) in the 1970s. The PSE follows fluctuations in the frequency modulations of the voice, tracing them on a chart to reveal when a person is lying. By adapting this equipment, plant researchers have been able to accomplish the remarkable: they have captured the voice of a plant.

Plant speech sounds like the hum of high-voltage wires, with a rhythmic, songlike quality to it. This is interesting, since shamans have spoken for thousands of years about the "song of the universe"—the energy vibration beneath all things. On cool desert nights, when you remove yourself from civilization and enter into the natural world, this vibration is audible to the human ear. The Andean shaman Domano Hetaka describes it in this way: "Life is conscious, intelligent energy. . . . It sings to everyone around it of its existence. This is life energy being scattered out as a gift from the heart."[14]

During ceremonies with visionary plants, this hum, this song, is also audible. It is as if the plants act as the antennae for the great music of the spheres and translate its language into bodily experiences.

It has been shown, using PSE techniques, that when a human being speaks to a plant, the plant responds in its own language. A conversation ensues. By communicating with plants in this way, researchers have found evidence for a humanlike intelligence. Plants can, for example, count. In answer to the question "What do two and two make?" a plant connected to PSE equipment will produce, on a chart, ink tracings with four peaks.

The Shipibo shamans of the Amazon know this plant language and intelligence well. They understand that the universe is composed of sound and it is the same sound as the rainforest and its plants. They can even capture it pictorially by transcribing its vibrations into geometric designs that they use to adorn their fabrics and craftwork. This work is so extraordinary that it is possible to enter a trance state by just looking at the intricate and complex designs—and all of this is the language of the plants that swells beneath the visible world and holds the universe together.

An example of the Shipibo mastery of plant language and the connection between sound and form is in their creation of huge clay pots that carry these geometric sound designs. Many pots are larger than a human being, and two Shipibo women are required to paint each one (only women do the craft and painting work). The women sit on opposite sides so they cannot see what the other is painting, but all the time as they work they whistle or sing the same chant, which is the song of the universe and the plants. When the painting is complete, no matter how complex the design, both sides of the pot always match perfectly. The women represent the plant-songs they chant in an exact three-dimensional form by speaking to each other in a universal language.

Of course, plants talk not just to us, but to each other. Experiments in England by Dr. A. R. Bailey show that when one of two plants in a greenhouse is watered, there are voltage changes in both plants. "There was no electrical connection between them, no physical connection whatsoever, but somehow, one plant picked up what was going on with the other."[15] *How* they picked this up is through what we would call language.

It may also be that plants can transmit more than "words." Soviet researchers have found, for example, in a step beyond Bailey's work, that a plant given water can, through some form of mysterious transfer, share it with another—as if all plants are connected in some way. In one experiment, a cornstalk in a glass container was denied water for weeks. It continued to live, however, and remained as healthy as others nearby. It was as if, said the researchers, the plant in the container had been "fed" by its neighbors, who shared their water with it. Such cooperation is one of the fundamental values we notice in nature and particularly in plants.

WHAT SHAMANS KNOW

Mojo bags, offerendas, and divinations, where plants are not ingested, still produce healings and change our realities. The evidence we have looked at suggests why. It is because plants have intelligence, awareness, and an affinity with human beings. There is communication between us, either spoken aloud or through the magic of the mind used with intention: whatever we think or ask of a plant, there is a response to our call, even over considerable distances. If a plant can hear and respond to the thoughts of someone seven hundred miles away, why should there be any need for us to ingest it for it to act on our healing intent?

Whenever we hold a mojo bag charged with our intention for love and luck, or make an offering to the spirits to receive their blessings, this is exactly what we receive, since the nature of nature is cooperation. It *wants* to share with us. Plants respond to and are motivated by love. Indeed, the healing nature of the universe *is* love, and plants are doorways to this, allowing us to tap into the greatest source of power for our emotional and physical well-being. And it is the pleasure of the plants to serve in this way.

Actually, they know no other way. As Vogel remarked, "Plants have always been my faithful companions and their wonderful healing properties have never once let me down. I am utterly convinced that nature provides us with everything we need to protect and maintain our health."[16] To explore this further, try the following exercises.

Making an Offering

An offerenda is an acknowledgment of the unity between humanity and nature and of respect between all of the species. It is also recognition of the fact that we cannot just take from nature without giving back. There is an ethical consideration here, of course, but even in practical terms we would not expect to keep using a battery without having to recharge it, and so it is with nature. Thus, in Haiti, after the medsen fey picks his leaves, he reciprocates by leaving a payment for the spirits, and in the Andes, the shaman makes an offering first before receiving spiritual favors.

To make an Andean-style offerenda you will need a piece of paper (approximately twelve inches square), one of the leaves or roots from your plant ally, and one or more "delicacies"—gifts that you think the spirits would like—such as candies, corn, peanuts, herbs, coins, and so on: whatever feels right to you. Each of these should be as dry as possible, so do not use moist herbs or make liquid offerings. You will also need a lighter or box of matches.

Take all of these items to a place in nature (a high place such as a mountain or hillside is preferable, but anywhere you will not be disturbed is fine) and spend a little time tuning in to your surroundings, gathering your intention, and making contact with the spirits of this place.

When you are ready, arrange the gifts you have brought on the paper and fold it once across the middle, then fold over the edges on each of the three open sides so you create a package. Hold this above you as you kneel on the ground, and state your prayers and requests out loud so the universe will hear you and bring you the blessings and good fortune you ask for.

When you have finished stating your needs, say: "Look, I have brought this gift for you, to express my gratitude and so my prayers will be answered" (or words to that effect). Then set fire to the package. After it has burned up, take the ashes and remnants and bury them, either there at the offering site or at the roots of the first tree you come to on your way back down the hillside. Do not look back, and remain in silence until you are home.

Your ally is a part of this offering and acts on your behalf as an intermediary with the spirits, so send thanks to your ally as well for the sacrifice it has made on your behalf.

Making Mojo

A mojo bag is an offering you carry with you. Or rather, it is like a spiritual radio, transmitting its energy and your requests to the universe and receiving the blessings that the spirits send back, through your attunement to these energies. To make one, you will require a piece of fabric (approximately seven to nine inches square), ribbon (a twelve-inch length is plenty), and, of course, the items that will go inside.

How you choose the color of this fabric and the things you include is determined by *your* needs and *their* signatures. Some colors in our culture are so archetypally associated with certain energies that we almost don't even need to think about them. We know that red is a symbol of power, that pink is for love, green is for healing, blue for peace, white for purity, and so on. Choose the color of your fabric according to which of these qualities you want to draw into your life.

You must also determine the contents of the bag by similar considerations. Paracelsus (see chapter 1) noted that the Creator had given us clues to the healing and spiritual powers of the plants by ensuring that their outward appearances reflect their use and intentions. Thus, the seed pods of the honesty plant, dried and relieved of their outer skins, are shining, translucent, and silver, round like coins and round like haloes. They can therefore be used to attract spiritual blessings and truth (honesty) or to draw in wealth, according to your intent.

Before you create a mojo bag, be clear on your intention (what you want the charm to do). Holding this intention in mind, take a walk in nature and allow yourself to be drawn to the plants that reflect your purpose. Spend a little time with them and speak of your needs. Request that they offer a little of themselves to help you.

When you take them home, dry and then desiccate them. (Backster and Vogel both found that plants retain their energy and intelligence even when dried and shredded. As long as the connection to a human being had been established, a galvanometer would still produce a reaction from a desiccated leaf.) Then mix them together with a little rum and Florida Water (or perfume), allow them to dry again, and wrap them in the fabric. You can add other items according to your needs—a few small coins, for example, to attract bigger money on the homeopathic principle of like attracting like; a rose quartz for love; a magnet to pull in success; and so on. You can also write out your prayers on a piece of paper and include them if you wish.

When you are ready, sew up or tie the bag with ribbon to finish it, then purify and consecrate it in incense or the sage smoke of a smudge stick, and sprinkle it with a little Florida Water or perfume. Carry it with you for good luck.

THE WILEY EXPERIMENT

Recall Vogel's student, Vivian Wiley, who conducted an experiment where she picked two leaves from a saxifrage plant and for two months projected love toward one and ignored the other. At the end of this time, the leaf that was loved remained fresh, despite being out of water, while the other crumbled and died.

As an interesting update on this, Jeremy Narby, in his 2005 book *Intelligence in Nature*,[17] quotes Dr. Anthony Trewavas, a scientist at Edinburgh University in the UK, who has been working with genetically modified tobacco plants for a number of years. One of the things that amazed Dr. Trewavas was his discovery that these plants respond immediately to touch. Even though they have never been regarded as touch sensitive, "one gentle stroke caused the modified plants to glow," says Narby.

The reaction of the plants was "as fast as we could measure," Trewavas commented. "Whereas I have been telling you that plants only respond in terms of weeks and months, in this case, they were responding in milliseconds to a signal which we knew would later have a morphological effect. If you keep touching a plant, it slows down its growth and it gets thicker."

Experiments like Wiley's are interesting because they are simple to perform and because they demonstrate the power of love and intention—the energies behind all shamanic healing with plants. But it's not necessary to act cruelly toward plants in order to validate these results. As a variant, you could try the following.

The Effects of Loving Intent

Purchase two identical potted plants—the same species, of course, and as close as you can get to the same age, height, size, and so on—and place them in different areas of your home, while ensuring that the conditions (light, heat, airflow, etc.) are as similar as possible. Water and feed them together on a regular basis; there is no need to deny one of them for the purpose of this experiment.

Apart from this, however, emotionally ignore one of the plants (simply treat it as if it wasn't there) and spend at least fifteen minutes a day directing loving energy toward the other, with the intention that it will thrive and grow. Maintain this practice for two months. At the end of this period, measure both again and look for any differences between them. Is one bigger than the other? Is it healthier or more vibrant? The difference between them is the measure of your love and intention.

When the experiment is over, explain to the love-deprived plant what you have been doing and why. Make it a promise that from now on you will send it love in equal measure. And, of course, you must keep your promise.

2

THE SHAMAN'S DIET: LISTENING TO THE PLANTS

*Whether the diet is to heal the body or spirit, or whether
it is part of an apprenticeship, what makes it work is your
good intention toward it as well as the intention to make
a connection to the spirit of the plant.
When the spirit accepts the dieter and the dieter has
the will, the spirit will grant him energy. The path to
knowledge opens and the healing can take place.*
GUILLERMO AREVALO, AMAZONIAN SHAMAN

This chapter will open you further to the spirit of the plants through
a process known as the *shaman's diet*. This body of practices involves
certain actions and restrictions on the behavior of the shaman-to-be so
he (or she) can learn from his plant ally how to use it for healing, and
how to strengthen himself physically, psychologically, and spiritually. In
Peru, it also includes the learning of magical chants called *icaros* and in
Haiti, the memorizing of sacred songs called *chantes,* to call and make
contact with the spirits and to invoke the power of the plants. All of
this is known as the *diet,* the purpose of which is to prepare the body
and nervous system of the apprentice for the expansion of consciousness
brought by his communion with the spirit of the plants.

The apprentice needs to follow dietary requirements in order to allow this transmission of healing power and knowledge. These requirements prohibit certain foods, such as pork, fats, salt, spices, condiments, and alcohol, leaving the apprentice with a healthy but extremely bland menu so she is not overwhelmed with flavor and can more finely sense the attributes of the plant she is working with.

The further purpose of such an uninspiring menu is to weaken the emotional attachments to everyday life, some of which come from food. For the same reason, there is a prohibition on sexual activity and the apprentice must even refrain from libidinous thoughts, since sex is another distraction that will ground him in his body and inhibit his spiritual potential. Detachment, and often removal, from the physical world, is, in this sense, a prerequisite to entering the great plant mind.

In the words of Guillermo Arevalo, a Shipibo *moraya* (the highest level of plant maestro):

> Above all, sex must be avoided or you will suffer a clash of energies. Sex debilitates a person and can produce *cutipa*—a psychological "accident"—in him. This means that the plant, instead of becoming a medicine, becomes a toxin and can provoke allergy, poison the blood, or cause heart problems. In these circumstances, it is said that the plant becomes jealous of the human lover and can make you ill or kill you. That is why the shaman goes into the wilderness. There is no temptation there.

A person who embarks on such a diet will also follow a regime of purification and retreat, which can last for weeks or months, and provides the quietude and focus for him to release his hold on the habits of normal life and still the rational mind so his spirit can expand. One example of this is when a male dieter is in his *tambo* (a small isolated shelter or meditation chamber, often deep in the jungle) and an old and unattractive woman is chosen to bring food to him. This prevents the dieter from entering into the sexual games-playing and social expectations that go with male-female interactions in the wider world and enables him to focus on his inner work.

According to Schultes and Winkelman, "Diet is viewed as a tool helping to maintain the altered state of consciousness which permits the plant teacher to instruct, provide knowledge, and enable the initiate to acquire power. The diet is viewed as a means of making the mind operate differently, providing access to wisdom and lucid dreams."[1] From a psychological perspective, it is a way of releasing the hold of the ego-mind. As the diet progresses, even the sense of being human diminishes, and the dieter becomes "plantlike." It is under these conditions that he or she can best start to communicate with the plant in dreams and meditations.

THE CHALLENGE OF THE DIET

The biggest challenge for a Westerner undertaking this diet is often not the requirements and prohibitions of the regime itself, but to accept that there is another order of nonmaterial reality that the apprentice can experience through his entrance into plant consciousness. We are all born into the social paradigm that surrounds us, with all of its beliefs, myths, and institutions that support its view of the world, and it is not within our own worldview to accept the immaterial and irrational. Before we, as Westerners, embark on such a plant diet, then, we first need to question some of our most deeply ingrained assumptions. The starting place for the diet is ourselves.

HomoSapien-centricity is a strange-looking word but perhaps an appropriate one to describe the concept that many of us, consciously or not, carry within us: that we humans are the most important (and even the *only*) conscious and self-aware—that is, ensouled—beings in the universe.

At his trial for heresy in 1633, Galileo, the Italian physicist and astronomer, was forced by the Inquisition to recant his suggestion that the Earth revolved around the Sun and not the other way around, as the dogma of the times dictated. It took nearly 360 years for the Church to pardon Galileo and accept his theories. Yet the arrogance behind the Church's judgment remains today, in the still prevalent notion that human beings are the "Crown of Creation" or, more kindly but just as arrogantly, that we are the "caretakers" of the natural world. In fact we are *members* of the world, equal parts of it; children of the Earth. By trying to control it or caretake it, we distance ourselves from the very Creation we claim ourselves owners of.

Another aspect of this HomoSapien-centricity is a science of psychology that regards nonordinary or expanded awareness as an extension or invention of the human psyche—perhaps even a delusion, but certainly not an experience of a real and bigger consciousness that can possibly exist without us. The soul of the world and the spirit of nature are not real in this model. It is we who create them.

Perhaps the closest we get to an ensouled view of the world is the concept of archetypes, which some psychologists regard as energies or spiritlike entities that may be external to ourselves. But even these are expressions of human consciousness and projections of *our* collective psyche, not beings in their own right.*

Is there an alternate reality that exists alongside the one we know? Another way of getting at this is to ask if time and space—our most basic building blocks of reality—are all there is. Is our perception of time correct: that it is linear and sequential, like a river flowing only from the present to the future? Is space made up only of material things and the gaps between them? Or do we create our own times and spaces, moment by moment, from the data available in the multiple dimensions around us? Our scientists who are now basing their theories on the existence of ten dimensions believe so. In the new science, in fact, this is the only way reality works.

Ultimately, it comes down to a matter of choice and personal belief. As John Michell comments in *Confessions of a Radical Traditionalist:* "There are an infinite number of ways in which you can see the world and an infinite range of data to support, or discredit, any of them. You

---

*Jung himself, however—ironically, the pioneer of the concept of archetypes and the collective unconscious—thought differently than the psychologists who interpret his work. In his *Mysterium Conjunctionis* he writes, for example, "It may well be a prejudice to restrict the psyche to being 'inside the body'. In so far as the psyche has a nonspatial aspect, there may be a psyche outside the body region so utterly different from 'my' psychic sphere that one has to get out of oneself . . . to get there."

can believe in black holes if you like, or you can believe in angels. I am not a believer, but if I had to choose I would take the latter, because unlike the holes, angels have often been sighted and their influence has generally been for the good."[2]

For shamans, the world we perceive through our senses is just one description of a vast and mysterious unseen, and not an absolute fact. Black Elk, the Oglala Sioux medicine man, remarked (in words that seem almost Platonic) that beyond our perceptions is "the world where there is nothing but the spirits of all things. [It] is behind this one, and everything we see here is something like a shadow from that world."[3]

In his book, *People of the Sacred Waterfalls*, Michael Harner writes of a similar view among the Jivaro people of the Ecuadorian rainforests. For them, "the normal waking life is explicitly viewed as 'false' or 'a lie', and it is firmly believed that truth about causality is to be found by entering the supernatural world or what the Jivaro view as the 'real world', for they feel that the events which take place within it underlie and are the basis for many of the surface manifestations and mysteries of daily life."[4]

Mazatec shaman, Maria Sabina, said the same thing: "There is a world beyond ours, a world that is far away, nearby and invisible, and there is where God lives, where the dead live, the spirits and the saints. A world where everything has already happened and everything is known. That world talks. It has a language of its own."[5] And the way that it talks is through nature. Though it is the antithesis of our social conditioning, this is the way to meet the spirit of the plants.

Loulou Prince reveals more about how this communication works when he speaks of the importance of dreams and of respect for the plants:

> Very often if a patient comes to me I have to send them away on that day. Then I will fast and prepare myself so that I invite a dream that night. In my dream the spirits come to me and lead me into the forests where they show me the leaves to pick and how to prepare them. Then, when my patient comes back next day I know exactly what I must do.

I must also pay the spirit for the help it has given me. At every crossroads I pass on my way into the forest to pick the leaves I have been shown, I must sing to the spirits and bury coins in the earth for a safe passage. The value of the coins is irrelevant; it is respect that is important.

When I reach the plant I will also sing, to wake up the spirits of the leaves and remind them of the dream we shared and the task that lies ahead.

Once the leaves are collected they must not touch the ground either for that would be an insult to the spirit and the energy of the plant would drain into the Earth.

The voice of nature is a dream, and the identification and collection of leaves is an exercise in respect. It begins with Loulou following a particular diet (often fasting) to summon the spirit to a dream-appointment. He will then follow the advice of this spirit and collect the leaves he is shown, all the time paying the spirits of the plants and the others along his route. Only after all this is done can the healing begin.

There are few illnesses that cannot be healed with the plants of the rainforest once these rituals have been followed. Loulou treats people with digestive disorders, sexual problems, fevers, and colds, and also has medicine to purify the spirit and restore balance to the body. He often treats children who are not growing well due to persecution by evil spirits. Here, the medicine is magical in nature.

There are specific leaves, strong-smelling leaves, that help children under spiritual attack. I collect the leaves I am shown by the spirits and mix them with rum and sea water to make a bath for the child, then I pray and bless the leaves.

While I am bathing him, I sing songs for the spirits, and ask them to come and help this child. Then I give him leaves to make his blood bitter, so it tastes and smells bad and the spirits go away. When this is done, no one can curse that child or do evil to him.

PLANTAS MAESTRAS:
THE SHAMAN'S TEACHERS

Plantas maestras ("master" or "teacher" plants) are an integral part of every diet. They are key among the shaman's tutelary spirits, her chief allies and guides to the worlds of health and healing. In ordinary reality, they are also considered the jungle's most skilled and important "doctors" because of their usefulness and relevance to the healing concerns of most patients. Through knowing these plants, the shaman can deal effectively with the diseases of her people.

It can be difficult to find discrete Western analogues for some of these jungle plants, because plants grow where they are needed—and the healing needed by a New York banker will be quite different from that of a Peruvian farmer. The psychological and spiritual benefits bestowed by such plantas maestras, and their ability to restore emotional balance, banish negative energies, or open the heart to love are desirable in any culture, however, so it is possible to find plants with equivalent or similar effects if we wish to imbibe these qualities for ourselves.

With this in mind, we now offer a description of some of the more commonly "dieted" plantas maestras of the Amazon along with plants from our own culture that will produce similar effects (either singly or in combination). (Also see appendices 1 and 2 for more Caribbean and Peruvian plants and their Western equivalents.)

Before we begin, let us clarify our terms and conditions. *To diet* the plants means to ingest them on a regular and sustained basis with the intention to merge with their spirit. If you choose to diet these plants, two things are essential.

First: That you take the plants at least twice a day for a period of not less than three months. Shamans typically spend many months or even years dieting a particular plant, so three months should really be considered a minimum.

During this time, the rest of your diet should be as bland as possible (rice, fish, and unseasoned vegetables are recommended) and you should avoid other herbs and spices as much as you can, as well as alcohol and stimulants such as coffee. Especially, during your diet, do not eat lemons or limes, which cut through magic. (In fact, the classic way for a shaman to end his diet is to eat a little salt and lemon or take a lemon bath.)

Second: That you focus on your intention. As you work with your chosen plants you should vigorously *intend* that you will meet with their spirits, not just ingest them for their physical properties. This means taking them with full awareness that you are allowing into your body a whole other intelligence.

We are not used to this discipline in the West. For example, we may be partial to mint and cumin, enjoying their tastes but blind to their effects on our body—how mint cools and cumin heats us up. When you embark on the diet, however, such subtleties of awareness are important, so begin to scan your body and emotions before and after you take your plant. How do you feel emotionally? What is going on physically in your body? What is your mind doing (are you stressed out or relaxed, etc.)? Where are you at spiritually? Then be aware of what changes—immediately and a few hours after you take the plant. In this way you will begin to sense its spirit.

To make a tea of any of the plants listed below, simply boil the fresh ingredients (the amounts you use can be much to your own taste, but three heaping teaspoons of each is about right) in a pint or so of water for a few minutes and then simmer for about twenty minutes, allowing it to reduce, and blowing smoke—which carries your intention—into the mixture as it boils. This will wake up the spirits of the plants and attune them to your needs. Add honey if you wish, then strain and drink when cool.

For a mixture that will last a little longer, add the fresh ingredients to alcohol (rum or vodka is recommended), with honey if you wish. Leave a little time for the herbs to soak into the alcohol,* then drink three to five teaspoonfuls a day, morning and night.

Here, then, are the plant teachers that Amazonian shamans recom-

---

*The length of time for this maceration will depend on factors such as ambient heat and exposure to sunlight, and so forth. In a sunny climate 8–10 days should be fine; in a cooler climate, three weeks or more is recommended.

mend, along with equivalent European and American herbs that will produce a similar effect.

CHIRIC SANANGO: FOR LOVE

Chiric sanango grows mainly in the upper Amazon and in a few *restingas* (high grounds that never flood). It is good for colds and arthritis and has the effect of heating up the body. (*Chiric*, in Quechua, means "tickling" or "itchy," an allusion to the prickly heat it generates.) Plant shamans often prescribe it for fishermen and loggers, for example, because they spend so much time in the water and are prone to colds and arthritis. The patient should not drink too much at a time, though, because it can lead to numbness of the mouth as well as a feeling of slight disorientation. It is also used in magical baths (see chapter 6) to change the bather's energy and bring good luck to his ventures.

Used in the West, the plant has a more psychological effect, but still having to do with "heat." Here, it enables people to open their hearts to love (it "warms up" a cold heart, but will also "cool" a heart that is too inflamed with jealousy and rage) and to identify with others as if they were brothers and sisters. In essence, it helps people get in touch with the sensitive and loving parts of themselves. Another of its gifts is enhanced self-esteem, which develops from this more healthy connection to the self.

Chiric sanango can be prepared in water or in *aguardiente* (weak sugar cane alcohol) or made into a syrup by adding its juice to honey or molasses. It is said to better penetrate the bones, however, if it is simply boiled in water and drunk.

For a Western diet, mint, as a balancer of the body's physical and emotional heat, has some of the properties of chiric sanango. It can cool you down on a summer day but will also provide warmth when drank by an open fire in winter; and it has the same effect on the emotions, promoting the flow of love as well as alertness and clarity. For these reasons, mint is associated with the planet Venus, which was named after the Roman goddess of love.

Good plants to combine with mint include lemon balm and chamomile. In Arabian herb magic, lemon balm was known to bring feelings of love and healing. (Pliny remarked that its powers of healing were so great that, rubbed on a sword that inflicted a wound, it would staunch the flow of blood in the injured person without need for any physical contact.) Recent research at Northumbria University in the UK has also proven the beneficial effects of lemon balm in increasing feelings of calm and well-being, as well as in improving memory. Chamomile, meanwhile, is a great relaxant and an aid to practice in meditation and forgiveness.

Chiric sanango, as mentioned, also brings relief from arthritic pain. If this is your concern, Western plants that could be added to mint include marigold and ginseng.

GUAYUSA: FOR LUCID DREAMS

Guayusa has the effect of giving lucid dreams (dreams during which you are aware that you are dreaming and can direct the dreaming). For this reason it is known in the Amazon as the "night watchman's plant," as even when you are sleeping you have an awareness of your outer physical surroundings. The boundary between sleeping and wakefulness becomes more fluid and dreams become more colorful, richer, and more potent than before. For those interested in dreams or shamanic dreaming, this is the plant to explore.

Guayusa is a good plant for people who suffer from excessive acidity, indigestion, or other problems of the stomach and bile. It also develops mental strength and is paradoxical in the sense that, just as chiric sanango is cooling and warming at the same time, guayusa is both energizing and relaxing.

In the Western world, bracken, jasmine, marigold, rose, mugwort, and poplar will produce the same effect of lucid or prophetic dreams. The leaves and buds of the poplar were often a key ingredient in the "flying ointments" of European witches, who used it for what we would call astral projection. Use a mixture of these plants (in either water or alcohol) to produce a liquid that can be taken in the same way as the examples above.

As well as preparing these plants as a tea or tincture (in either water or alcohol) it is also possible to use them in a way that practitioners of Haitian Vodou employ for working with their native "dreaming plants": by making a *bila*. In Haiti, the ceremony of *bat guerre*—the "battle for the spirits"—is an initiatory ritual that takes three nights to complete and is undertaken by the person who wishes to become a shaman. The initiate kneels before the bila, a ceremonial pillow that is stuffed with magical herbs, and, for hours on end, must beat it with a machete, while drumming and dancing carry on around him. This beating releases the aroma of the herbs, and through this, he will enter a trance in which the spirits talk to—and, often, through—him.

The bat geurre takes place before the entire community, since it is also about demonstrating commitment to a spiritual vision of the world. The sweat of the new priest and his blistered hands are proof to the people that he has the dedication and strength to make this vision a reality on Earth.

Thankfully, it is not necessary for you to undergo this ritual in order to sense a new reality, since we all enter the dreaming universe every night, when we sleep. Many people understand that herbs can influence them subtly and may sleep with a pouch of lavender next to their beds, or with rose- or primrose-scented sheets. They find that this relaxes them and changes their mood in slight, but perceptible, ways. The herbs of the bila are, in this respect, no different.

To create a bila for yourself, take small handfuls of mugwort and poplar, or some of the other herbs mentioned, and blend them together. Sprinkle the mix with neroli, orange, or patchouli oils (aromatherapy oils are fine) as well if you wish and, as they do in Haiti, a little rum and water to bind the mix together. Put your intention into this as you do it: these herbs *will* help you dream more lucidly and gather information from the spirit world. Then allow the mixture to dry for a few days. When it is ready, crumble it into a cloth pouch and place it beneath your pillow. Keep a dream journal next to your bed, and as soon as you wake up in the morning, immediately note down your dreams and your first waking sensations.

Julie, a participant in one of our 2003 plant medicine workshops, made and used a bila in this way. Not only was her dreaming enhanced tremendously and the information she recorded in her dream diaries richer than she had experienced for many years, she also reported out-of-body sensations that she had never had before.

"It felt like I was flying over landscapes during my dreams," she said, "and I could zero in on any location and explore what was going on there. Quite often these experiences would involve friends and when I would later call or meet with them it turned out that, more often than not, what I had seen in my dreams had actually happened to them. In one or two cases they also experienced my presence in their own dreams on the night I was dreaming of them."

A similar method for inducing dreams that will impart specific information or healing comes from the Hindu tradition, which uses a dream pillow. The herbs that you use in this tradition are determined by which facets of your life you need guidance with. The list below offers some correspondences to the spices, herbs, and flowers that dreamers find useful.

First, make a base mix of mugwort and poplar, as described for the bila. Then, for information on the following, add the corresponding herb.

- Love: Rose and/or cinnamon
- Money: Bergamot
- Health: Fennel or lavender
- Stopping nightmares: Cloves
- Spiritual development: Jasmine

Cut a piece of natural cloth (silk or cotton is fine) into a rectangular shape about twice the width of a CD case. Black or white fabrics are normally used, but you can choose colors appropriate to your needs or tastes. Fold the rectangle in half so it forms a square, then sew up the two sides and turn the sleeve you've created inside out. Fill the pillow with your "base" herbs, adding the flowers and herbs most appropriate for the guidance you want to receive. Sew up the top and you will have a dreaming pillow.

It is said that an intention for dreams based on love is best made on a waxing moon (the fourteen days starting from the new moon) and dreams about health are best on a waning moon (the fourteen days following the full moon).

AJO SACHA: STALKING THE SELF

Ajo sacha is a blood purifier that helps the body rid itself of toxins (spiritual or physical) as well as restoring strength and equilibrium lost through illnesses that have an effect on the blood. (See a photo of ajo sacha on page 2 of the color insert.)

Psychospiritually, ajo sacha helps to develop acuity of mind and can also take the user out of *saladera* (a run of bad luck, inertia, or a sense of not living life to the full). It is also used for "ridding spells"—undoing the work of curses or removing bad energy that has been sent deliberately or by accident (in an explosion of rage, etc).

In floral baths (see chapter 6), it will relieve states of shock and fear (known as *manchiari*). These states can be particularly debilitating to children, whose souls are not as strong or fixed as an adult's; a powerful shock can therefore lead to soul loss (see chapter 4). The same phenomenon, especially regarding children, is known to the shamans of Haiti, where it is called *seziman*, and the shamans of India, where great care is taken to protect children from frights of this kind and the anxious parents of newborns employ the shamans to make protective amulets for their babies.

Another key use for ajo sacha in the Amazon is to enhance hunting skills, not only by covering the human scent with its own garlicky smell (the plant also has a strong garlic taste and, in fact, its name—a hybrid of Spanish and Quecha—means "wild garlic," although it is not related to garlic in any way), but by amplifying the hunter's senses of taste, smell, sound, and vision, all of which are, of course, essential for success and for survival. It is *a plant of stalking*.

In the Western world this stalking ability tends to translate psychologically, and the plant becomes a means of helping an individual hunt or "stalk" her inner issues. To underline this, the Shipibo maestro Guillermo Arevalo adds that this plant opens up the shamanic path and helps the apprentice see beyond conventional reality—as long as he has "the heart of a warrior" and is prepared to live under the obligations of shamanism. For this, he will need courage, the ability to face the truth, and to know his true calling without fear of extremes or "ugly things."

It is fascinating that this plant, used for hunting in the rainforest, possesses the same essential quality in an environment such as ours where food is purchased from supermarkets and we do not track down game at all—but we often *do* have work to do in stalking ourselves. Clearly, this plant has extraordinary qualities.

Western plants with equivalent therapeutic uses include valerian and vervain. Valerian has been recorded from the sixteenth century as an aid to a restful mind and, in the two world wars, was used to combat anxiety and depression. Today, we still use it for these purposes. It also brings relief from panic attacks and tension headaches, which are regarded as symptoms of an underlying cause—an unresolved issue or stress of some kind. By relaxing the mind, valerian enables the psyche to go to work on the real problem, aided by the plant itself.

One way of dieting valerian (which will also aid a deep and restful sleep) is by adding equal parts to passionflower leaves and hop flowers and covering it with vodka and honey for a few weeks, after which a few teaspoons are taken at bedtime.

Vervain, meanwhile, was well known to the druids, who used it to protect against "evil spirits" (nowadays, we might say "inner issues" or the "shadow-self"). We use it to help with nervous exhaustion, paranoia, insomnia, and depression. Like valerian, by relaxing the conscious mind it empowers the unconscious to go to work on—or stalk—the more deep-rooted problem.

Garlic is another protective plant with the effect of purifying and strengthening the blood. Nicholas Culpepper noted its balancing qualities and wrote of it as a "cure-all." It has long been associated with magical uses, protection from witches, vampires, and evil spells, and as effective in exorcisms (i.e., psychologically speaking, in ridding us of our inner demons). Roman soldiers ate it to gain courage and overcome their fears before battle. There is also a tradition of placing garlic beneath the pillows of children to protect them while they sleep and defend them from nightmares.

One way of dieting garlic is in the form of garlic honey—which is not as disagreeable as it might sound. To make it, add two cloves of peeled garlic to a little honey and crush them in a mortar, then add another 400 grams or so of honey to the mix. Drink this in hot water or simply eat it, two teaspoons a day, morning and night.

Other plants that are good for increasing wisdom (inner knowledge) include peach, sage, and sunflower. All of these can be dieted fresh or in a little rum or vodka.

MOCURA:
PSYCHOLOGICAL AND EMOTIONAL STRENGTH

One of the qualities of *mocura* is its ability to boost one's psycho-emotional powers and bring equilibrium. For this reason it is regarded as a great balancer, restoring the connection between the rational mind and the feeling self. For example, it is good at countering shyness and can enhance one's sense of personal value and authority by helping overcome painful memories (of past embarrassments and failures, etc.).

Mocura is also used in floral baths to cleanse and protect against malevolent forces such as sorcery and *envidia* (envy). Its medicinal properties include relief from asthma, bronchitis, and the reduction of fat and cholesterol.

In the West, there are a number of plants that have similar effects, bringing calm and balance to the soul. These include lavender—which Pliny regarded as so powerful that even looking upon it brings peace—meadowsweet, pine, and rosemary.

Burning pine needles will purify the atmosphere of a house and a pine branch hung over the front door will bring harmony and joy to the home. Rosemary, especially when burned, is cleansing and centering, and it is said that if you concentrate on the smoke with a question in mind, it will provide the answer. There is a European belief that carrying rosemary will protect you from sadness, and it is quite pleasant to drink with honey as a weak tea.

In terms of body energetics and magical uses, moss, orange, and strawberry leaves are all used by European witches to remove bad luck, and loosestrife, myrtle, and violet leaves to overcome fear.

ROSA SISA: HARMONY AND HEALING THE SOUL

The *rosa sisa* plant is often used to heal children who are suffering from *mal aire* (bad air), a malady that can occur when a family member dies and leaves the child unhappy and sleepless. The spirit of the dead person lingers, it is said, because it is sad to go and aware of the grief around it, so it stays in the house and tries to comfort its family. This proximity to death, however, can make children ill.

Rosa sisa is also used to bring good luck and harmony in general. One of the ways that bad luck can manifest is through the magical force of envidia. A jealous neighbor might, for instance, throw a handful of graveyard dirt into your house to spread sadness and heavy feelings. Those in the house become bored, agitated, or restless as a consequence. The solution is to take a bucket of water and crushed rosa sisa flowers and thoroughly wash the floors to dispel the evil magic.

Many Peruvians also grow rosa sisa near the front door of their houses to absorb the negativity of people who pass by and look in enviously to see what possessions they have. The flowers turn black when this happens, but they go back to their normal color when the negative energy is dispersed through their roots to the Earth.

Rosa sisa is also used for making dreams come true, by blowing on the petals with a wish in mind, like we do with dandelions. It can make these wishes happen because it is "bright like the sun" and contains the energy of good fortune.

Rosa sisa are African marigolds *(Tagetes erecta),* and have similar magical uses in the West. Aemilius Macer, as long ago as the thirteenth century, wrote that merely gazing at the flowers will draw "wicked humours out of the head," "comfort the heart," and make "the sight bright and clean." In Europe, just as in Peru, marigolds are often grown beside the front door or hung in garlands to protect those inside from magical attacks. For the same reason, and to empower the spirit, marigold petals can be scattered beneath the bed (where they will also ensure good—and often prophetic—dreams) or added to bath water to bring calm and refreshment to the body and soul.

As well as marigold tea (which is good for bringing down fevers, especially in children), for gastritis, gallbladder problems, and tonsillitis, the petals can be dieted in salads or added to rice and beans. They can also be rubbed on the skin to heal irritations, cuts, bruises, and rashes.

Alternatives, to create harmony in the self and home, include gardenia, meadowsweet, and passionflower.

PIRI PIRI, MEDICINAL SEDGES: FOR VISION

Native people cultivate numerous varieties of medicinal sedges to treat a wide range of health problems. Sedge roots, for example, are used in Peru to treat headaches, fevers, cramps, dysentery, and wounds, as well as easing childbirth and protecting babies from illness.

Shipibo women cultivate special sedge varieties to improve their skills in weaving the magical tapestries that embody the spiritual universe. It is customary when a girl is very young for her mother to squeeze a few drops of sap from the *piri piri* seed into her eyes, to give her the ability to have visions of the designs she will make when she is older. The Shipibo men cultivate sedges to improve their hunting skills.

Since the sedges are used for such a wide range of conditions, their powers were once dismissed as superstition. Pharmacological research, however, has now revealed the presence of ergot alkaloids within these plants, which are known to have diverse effects on the body—from stimulation of the nervous system to the constriction of blood vessels. These alkaloids are responsible for the wide variety of sedge uses; however, they actually come not from the plant but from a fungus that infects it.

There are a number of Western plants that are also said to produce visions—that is, communion with the greater spirit of the world. The leaves of angelica and coltsfoot, when smoked, for example, will induce such visions; and damiana, when burned, will also produce these effects.

Angelica has long been regarded as a spiritual plant with almost supernatural powers. It is linked to the archangel Raphael, who appeared in the dreams of a medieval monk and revealed the plant as a cure for plague. Native Americans used it in compresses to cure painful swellings: it sucked the spirit of pain out of the body and cast it to the four winds. Angelica has been heralded as an aid to overcoming alcohol addiction, as its regular usage creates a dislike for the taste of alcohol. Recent research suggests that it can also help the body fight the spread of cancer. Its leaves can be added to salads; this is another way to diet this plant.

Coltsfoot is another plant with wide-ranging properties but is most highly regarded for its soothing effects on respiratory and bronchial problems. One way of dieting it, paradoxically, is to use it in herbal cigarettes. These can be made by adding a larger part of coltsfoot to other aromatic and soothing herbs such as skullcap or chamomile. Cut the herbs to small lengths and mix them thoroughly with a little honey dissolved in water, then spread the mix out to dry for a few days. You can then roll it to make cigarettes or smoke it in a pipe.

UNA DE GATO: FOR BALANCE

Una de gato (cat's claw) is a tropical vine that grows in the rainforests. It gets its name from the small thorns at the base of the leaves, which look like cat's claws and enable the vine to wind itself around trees, climbing to a height of up to 150 feet.

The inner bark of the vine has been used in the Amazon for generations to treat inflammations, colds, viral infections, arthritis, and tumors. It also has anti-inflammatory and blood-cleansing properties. It will clean out the entire intestinal tract to treat a wide array of digestive problems such as gastric ulcers, parasites, and dysentery.

Una de gato's most famous quality, however, is its powerful ability to boost the body's immune system, and it is considered by many shamans to be a "balancer," returning the body's functions to a healthy equilibrium.

From a psychospiritual or shamanic perspective, disease usually arises from a spiritual imbalance within the patient, causing him to become *de*-spirited or to lose heart (in the West we would call this depression). Interestingly, Thomas Bartram, in his *Encyclopedia of Herbal Medicine*, writes that in the West, "some psychiatrists believe [problems of the immune system, where the body attacks itself] to be a self-produced phenomenon due to an unresolved sense of guilt or dislike of self. . . . People who are happy at their home and work usually enjoy a robust immune system."[6] The psychiatric perspective, in this sense, is not so different from the shamanic view.

Cat's claw is believed to heal illness by restoring the peace of the spirit as well as the balance between spirit and body. The medicinal properties of this plant are officially recognized by the Peruvian government and it is a protected plant (in terms of export). It is, however, widely available in health food stores in the West, either in its natural form or as capsules, which are another way of dieting it, although its spiritual effects will be weaker, since, once a plant has been processed, much of its spirit is lost.

Using echinacea as a substitute for cat's claw will stimulate the immune system and prove effective against depression and exhaustion. As another alternative, you might try a mixture of borage, cinnamon, and blackberry, all of which are regarded as good healers in general and good for lifting the spirits.

CHULLACHAQUI CASPI: CONNECTION TO THE EARTH

The resin of the Amazonian *chullachaqui caspi* tree, extracted from the trunk in the same way as rubber from the rubber tree, can be used as a poultice or smeared directly onto wounds to heal deep cuts and stop hemorrhages. For skin problems such as psoriasis, the bark can be grated and boiled in water while the patient sits before it, covered with a blanket, to receive a steam bath. Oil can also be extracted by boiling the bark, and this can be made into capsules. It is important to remove the bark without killing the tree, however, which can otherwise have serious spiritual consequences.

The deeper, more spiritual purpose of this tree is to help the shaman or his patient get close to the spirit of the forest and in touch with the vibration and rhythm of the Earth. This reconnection with nature will strengthen an unsettled mind and help to ground a person who is disturbed. It will also guide and protect the apprentice shaman and show her how to recognize which plants can heal.

The tree has large buttress roots as it grows in sandy soil where roots cannot go deep. *Chulla* in Quechua, in fact, means "twisted foot" (a reference to the root structure, while *chaqui* is the plant). Amazonian mythology also discusses this and includes stories of the jungle dwarf, the chullachaqui, who is said to have a human appearance, with one exception: his own twisted foot. The chullachaqui is the protector of the animals and lives in places where the tree also grows. The legend is that if you are lost in the forest and meet a friend or family member, it is most likely the chullachaqui who has taken his or her form. He will be friendly and suggest going for a walk so he can guide you or show you something of interest. If you go, however, he will lead you deep into the rainforest until you are very lost indeed, and you will then suffer madness or become a chullachaqui yourself.

Perhaps this legend refers to the initiation of the plant shaman, who must go deep into the jungle to pursue his craft by getting to know the plants and the forest. Such trials can, indeed, lead to madness or even death for the unwary; but those who succeed will become great healers, in touch with the spirits of nature, like the chullachaqui himself.

The chullachaqui dwarf is also a symbol of the tree. The motif of the "world tree"—the spiritual center of the universe, which connects the material and immaterial planes—occurs in many cultures and is often connected with initiation. In Haiti, it is the spirit Papa Loko (whose name is a variant of *Iroco*, which is the name of an African tree) who meets the shaman-to-be in the dark woods at night to initiate him into the Vodou religion. In Siberia, too, there is a tradition that the shaman-elect must climb a silver birch while in a state of trance and make secret, spirit-given markings on one of its topmost branches.

For those who are not ready to meet the challenges of shamanic initiation, however, the advice of the jungle shamans is simple: when out walking in the forest, should you encounter a friend or a family member, always look at his feet, as the chullachaqui will try to keep his twisted foot away from you. Do not go with him—turn back and run away!

While it is interesting for us to speculate about the initiatory symbolism of the chullachaqui, Amazonian shamans regard it as a very real being. We have a photograph, given to us by Javier Arevalo, for example, which shows a chullachaqui's tambo, and Javier swears it is real. The tambo is a hut that stands about four feet high and is used as a jungle dwelling. Javier discovered this one next to a cultivated garden deep in the otherwise wild rainforest.

In the West, we have our own tradition of magical trees. One of these is willow, a tree sacred to the druids. Ancient British burial mounds and modern day cemeteries are both often lined with willow, symbolizing the gateway this tree provides between the living and the dead, spirit and matter. The brooms of witches are also bound with willow, enabling their flight to the otherworld.

To deepen a connection to the Earth and the spirit, willow can be dieted in place of chullachaqui caspi. Do this by burning crushed bark fragments with white sandalwood or myrrh and bathing in the smoke.

CHUCHUHUASI: INCREASED LIFE FORCE

Chuchuhuasi is another Amazonian tree that forms an important part of the jungle pharmacopoeia. The bark can be chewed as a remedy for stomachache, fevers, arthritis, poor circulation, and bronchial problems, but it is rather bitter and so more often it is macerated in aguardiente or boiled in water and honey.

Western alternatives include burdock for arthritis and for "fevers" that manifest through the skin in the form of eczema, psoriasis, acne, and so forth; and ginseng for problems of the circulation. Kola is good for stomach complaints (diarrhea and dysentery, etc.); and saw palmetto is a general tonic and useful for bronchial problems.

Chuchuhuasi is also regarded as a libido stimulant and aphrodisiac, giving the person who drinks it a renewed sense of life and vigor. With these properties in mind, chuchuhuasi is the main ingredient in cocktails at many bars and restaurants in Iquitos, on the banks of the Amazon river. The most popular of these is the Chuchuhuasi Sour, where it is mixed with lime, ice, and honey.

In the West, plants with similar aphrodisiac qualities include burdock, ginseng, kola, and saw palmetto berries. These are not just aids to sexual potency, but reconnect the dieter to the joy of living and a love of involvement with others.

At Ari's Bar in Iquitos, it is possible to buy many exotic and stimulating brews to help with matters of sexual potency. One of these is the aptly named Super Erectus, which is a blend of raw egg, boiled yohimbe bark, catuba, ginseng, guarana, kola nut, damiana, yogurt, fruit juice, honey, bee pollen, and crushed brazil nuts, cashews, and peanuts, all of which are mixed in a blender and drunk as a shake. The raunchy legend at Ari's is that one of his customers, an elderly man who drank two of these a day and was always in the company of women at least half his age, had to be buried in a coffin with a hole cut in the top when he finally died because, when he lay on his back, it was impossible to nail the lid shut. A story that speaks for itself.

In the strange jungle town of Iquitos, everyone is the author of ten thousand legends, so we should point out that this story has also been attributed to the shaman Augustin Rivas Vasquez, and in his telling of it, refers to the drink Rompe Calzon.[7] Since *rompe calzon* translates as "bust your trousers," however, the point is much the same.

If you would like to try a "super erectus" brew of your own, here is one of Ross' recipes, using herbs more commonly found in the West. Mix ginseng, saw palmetto, muira-pauama, sarsaparilla, helonia, agnus castus, kola, damiana, licorice, pollen, propolis, honey, and royal jelly to taste. You can then add this to rum or brandy for a tonic, or to raw egg and organic yogurt blended to make a shake. Drink a cupful each morning.

REFLECTIONS ON THE DIET

Dieting a plant *intelligence* is totally different than taking a pharmaceutical. The latter has an effect only while the course of drugs continues, whereas plant medicines lead to a permanent change through the relationship you establish with the essence of that plant. This connection, at first, may appear metaphorical or symbolic; but as it deepens, you will eventually experience it across all of your being: physically, spiritually, mentally, and emotionally as the spirit of the plant merges with your consciousness and begins to alter your psychospiritual or emotional DNA.

One of the great revelations (and comforts) that we can experience while working with plants in this way is that we are not separate from the natural world at all; we are all connected. Here, we have included a few exercises to enable you to work deeply with your plant allies and experience more of this connection for yourself.

A few obvious comments first:

1. Work with plants that grow locally.

   The healing needs of the various cultures will differ from one another. Sometimes this is a matter of climate and other physical factors (colds are more common in England than Peru, for example, because the climate is colder and damper). Sometimes it is a psychospiritual matter having to do with the inclination of a particular people toward a particular way of life. (Stress-related diseases are less frequent in Haiti and Jamaica than in the United States, for example, because the former are more "laid back," whereas people in the Western culture are more exposed to the rush, pressures, and backstabbing of the corporate world and the frenzy of modern life.)

Wherever we are, though, shamans tell us that the Creator knows and has met our healing needs and so local plants will always be stronger and more appropriate in our diets and cures. Many traditional ways and indigenous methods of working with the local plants have, of course, been lost in Western society; but the practices of the plant doctors of other cultures, which you have read about in this chapter, can be used just as effectively with our native plants. (Also check the appendices at the end of this book for other local plants that can be dieted for particular physical or spiritual needs.)

2. There is power in every part of a plant.

    Nothing need be discarded and we can learn from each flower, root, leaf, or fragment of bark. Even common plants (or so-called weeds) have spiritual and medicinal properties. It is not just the popular and pretty ones that we must always seek out for, as the sin eater, Adam, used to say: "A weed is simply a gift of nature that we do not care to receive," though its healing potency remains the same. Choose the plant that calls to you, irrespective of its status in the Western world.

3. The process of selecting a plant to diet is intuitive and emotional, not rational and cerebral.

    Your choice might result from many factors—the color or scent of a plant can be meaningful to you, or perhaps there was a flower you loved as a child and would like to know more about now. Just as you are "drawn" to someone who will become a new friend rather than sitting down and making a reasoned and objective assessment of whether you want them in your life, allow yourself to be drawn to your plant allies too.

How to Diet

As our example, let's take rosemary. Its distinctive scent is invigorating and stimulating, and maybe that sense of revitalization is a quality you want in your life. You feel that dieting this plant would help and you are drawn to it emotionally.

If you now "tune in" to or research this plant, you discover that rosemary has long been known as a blood and nervous system stimulant. Oil of rosemary is used in salves to treat rheumatism, nervous headaches, muscular aches, and sprains, for example, and when added to baths it helps tone the skin. Rosemary also has qualities of cleansing and purification and is used in traditional societies as an incense to cleanse negative and disharmonious places such as sickrooms.

So now you have an idea about the properties of this plant—all of which stemmed from your feeling that its scent was invigorating and stimulating. Your research supports this by showing that your emotional perceptions were right. So you decide to diet it.

Make Friends with the Plant

First, spend some time simply being with the plant. Look at it, noting its shape and colors, run your hands through its leaves, feel how smooth the body of each one is, but how sharp the tip, like a needle ready to inject its health-giving properties. Inhale its scent as you visualize its stimulating and purifying qualities entering your body. Take a leaf and taste it. Be playful and invite the plant to become your friend and teacher.

Gather the Plant

Before you pick any part of a plant, tune in to it again and it will tell you the best time for gathering it. Night gathering tends to infuse a plant with gentler and more "feminine" moon energy, for example, whereas picking at midday will mean it is charged with powerful and "masculine" sun energy. By the same token, gathering early in the growing season will give you a subtle, "adolescent" energy that is not yet fully developed (but that may be exactly what you want), while picking toward the end of the season—in the plant's "old age," as it were—will mean a plant filled with wisdom but whose energy is now returning to the soil as it begins its winter hibernation. Its knowledge will be great, in other words, though its power may be weaker. In midseason, the plant will be coming into wisdom and at its most powerful. There will always be an optimal time to gather, according to your needs, and the plant itself will reveal this. (Or you can, if you prefer, find a reference source in the form of an herbal encyclopedia that will give you some of the same information, though it won't tell you much about the spirit of the plant, of course.)

The leaves of rosemary, for example, can be gathered when they are fully developed but prior to the flowers appearing, as this best holds the power of the plant and retains the active ingredients in the leaf and stem cells.

Once you have taken what you need, air-dry the leaves, which you can then store in a moisture-sealed glass jar. Once they are dry, the active ingredients in the leaves will also be released more easily into water or alcohol.

Prepare the Plant

There a number of ways to prepare a plant, as we have seen. The easiest is to make an infusion. This is simply a tea made by steeping the leaves in freshly boiled water for ten minutes. As a guideline, use about an ounce (around 30 grams) of dried plant to two cups of boiled water, which will provide three doses of plant infusion. Increase the ratios if you want to make a larger batch that will last for several days.

Another method is to make a *macerado,* or tincture. Here, you macerate the leaves and stems in alcohol. (Vodka, which is pure and contains few other flavorings and colorings is probably best for this, but rum is also used.) Put the leaves and stems into a clean glass jar so they fill the container two-thirds full, then pour in the alcohol to fill the jar and seal it. Leave it in a cool, dark place for a few weeks, giving it a shake every other day. The advantage of this method is that the mixture will last for many months, so the plant is always available to diet.

Whichever you choose, remember that your *intention* is always the most important ingredient, so hold in mind your purpose for dieting the plant as you go through each stage of preparation. In this way, you reach out to the spirit of the plant and inform it of your needs.

Diet the Plant

Each morning before breakfast, take a half cup of the infusion or, if you have made a macerado, a half shot glass (about three teaspoons). Do the same in the evening. Find as much time to relax as you can while you do this, so you are undisturbed and can tune in to the plant.

After a week or so, you may start to find your life taking on some of the qualities of the plant itself. In the case of our example, as rosemary is stimulating, you might find that there is more going on around you, or that you have more "get up and go."

You may also find that your dreams become deeper and more meaningful. Or the spirit of the plant might appear to you in these dreams, either in the form of a person or as an event that has a mood or personality to it that is related to the characteristics of the plant. These things may also happen during meditation or shamanic journeying.

Keep up your diet for three months, and during this time, also bring fresh sprigs into your home, place leaves under your pillow, paint or draw the plant. As you maintain your practice, there will come a moment when you sense the plant actively reaching out to you. At that moment you will know that the plant is your ally—the door will be open for you to learn its ways, how it will help you, and how it can guide your deeper journey into the plant world.

Journey to the Plant Spirit

When you feel that this moment has come and the connection between you is strong, start your journey to the spirit of the plant (see chapter 1) so you begin to see it in whatever noncorporeal form it takes. Ask it to reveal more of itself and the great spirit of nature of which it is also a part.

3

PLANTS OF VISION:
SACRED HALLUCINOGENS

Ayahuasca is a shortcut. It's as if we had been traveling
down the same highway as the rest of humanity, but, in
order to arrive at our destination more quickly, we took
a side road . . . a shortcut that leads us to truth.
PADRINO ALEX POLARI DE ALVERGA

The reverence that indigenous people have for the natural world stems from their understanding that nature is not just physical but embodies spiritual realities. We in the Western culture have lost some of this fascination with nature, but it was not so long ago that we shared the awe of native people for this great and mysterious world of which we are a part. When Walt Whitman wrote the following words in "Song of Sunset," published in 1900, he was expressing the joy felt by any shaman (or anybody) who knows he is walking with spirit as he moves within the forest. How much we have lost in a hundred years.

How the clouds pass silently overhead!
How the earth darts on and on! And how the sun,
moon, stars, dart on and on!

How the water sports and sings! (Surely it is
  alive!)
How the trees rise and stand up—with strong
  trunks—with branches and leaves!
(Surely there is something more in each of the
  trees—some living Soul)
O amazement of things! Even the least particle!
O spirituality of things!

All children know this feeling too.

One of the traditional ways for the plant shaman to reaffirm his connection to this living and inspirited world and to allow it to communicate with and through him has been by use of visionary teacher plants, each of which is a spirit maestro or master of awareness in its own right. This is the domain of the sacred hallucinogens.

To a Westerner, the term *hallucinogenic* may mean only that these plants "produce hallucinations." This is a common, but rudimentary, view of their actual potential as well as of what hallucinations—and, by implication, what "reality"—might actually be.

For the shamans, we are all dreaming (or hallucinating) all of the time. Our modern cities and ways of life are the dreams of the West, embodying a myth of what the world is or should be. Fundamentally, the Western dream is one of separation and disconnection from the flow of things, where competition, conflict, and challenge are the norm. The fact that things do not *have* to be this way and we *could* create a different world based on richer, more inclusive, more liberating (or any other) principles—*but we do not*—suggests to the shamans a mass hallucinatory experience in its own right. We are so involved in the dream that we do not see an alternative to it.

Sacred hallucinogens are the means of breaking through this trance of the social dream into the expansive, freeing, information-rich universe full of infinite possibilities for other realities and futures. These plants do not lead us *away* from ourselves, into an unbalanced frame of mind, as our doctors and politicians warn, but deeper *into* ourselves and our potential; to a place where we can find greater balance through genuine self-awareness.

Sacred hallucinations are messages from spirit. They do not just deliver nonsense images of things that aren't there, as we might conceive of a hallucination, but offer the experience of a "true hallucination," to use a term from Terence McKenna, from which we see through the mists of socialization into our own possibilities and spirit.

The power of visionary plants requires that they always be taken in a ritual setting conducive to the appearance of the gods and with an intention or purpose in mind—for self-understanding and meaning-ful connection with a greater-than-human reality—and this setting and intention contribute to their effect as well.

The word *hallucination* might imply a primarily visual experience, but for the shaman it is more than that. Visions may come, of course, but teacher plants also bring with them an intense experience of ecstasy and oneness with the world, deep and profoundly meaningful insights, a searchlight on our hidden thoughts and feelings, through which our egos can let go and we can merge with a greater field of creative con-sciousness. It is the realizations, not the images, that are visionary. The hallucinogenic, as a total experience, offers a doorway into the hidden realms of human consciousness and the spiritual intelligence of a living planet.

SACRED EVOLUTION:
HALLUCINOGENS IN HUMAN CONSCIOUSNESS

The human brain shares an affinity with hallucinogens. Our neural chemistry contains some of the most powerful psychotropic compounds in the world, such as tryptamines and serotonin, which are identical to those found in many teacher plants. It is part of our design, our bio-logical blueprint, to be able to move into expanded awareness or deep trance almost at will, and this may be no accident of evolution.

Some, like McKenna, argue that our capacity for expanded con-sciousness and deep thought arose directly from the ingestion of plants such as fly agaric and psilocybin mushrooms back in the very early days when human beings were nomadic hunter-gatherers, barely human at all, who would forage for food and eat whatever they found.[1] Certainly it is true that a million and a half years ago, the human brain underwent what Rita Carter, in her book *Mapping the Mind,* describes as "an explosive enlargement."

> So sudden was it that the bones of the skull were pushed outward, creating the high, flat forehead and domed head that distinguish us from primates. The areas that expanded most are those concerned with thinking, planning, organising and communicating. . . .
>
> The frontal lobes of the brain duly expanded by some 40 percent to create large areas of new gray matter: the neocortex. This spurt was most dramatic at the very front, in what are known as the prefrontal lobes. These jut out from the front of the brain, and their development pushed the forehead and frontal dome of the head forward, reforming it to the shape of a modern skull.[2]

Nobody knows what caused this dramatic and sudden expansion of the brain, which separated us from the other animals and created the prototype for modern man. But, a sudden expansion of consciousness might do it—because we would need new grey matter in order to process and store visionary information downloaded from the plants.

If this is so, then for at least the last million and a half years, we have been hardwired for the sacred, even though many of us are largely denied it today. Indeed, so hungry are we for numinous experience and the freedom to truly use our minds in our true-hallucination-deprived world, that people are turning in increasing numbers to alcohol and drugs as their only available means of entrance to an alternate sense of reality. Unfortunately, many of these alternatives are addictive and deadening, in contrast to sacred hallucinogens; thus, taking them defeats the object of the quest. We have fallen from grace with the planetary mind, and we futilely reach for reconnection.

THE COSMIC SERPENT
AND THE VINE OF SOULS

One of the most potent and best-known of sacred hallucinogens is ayahuasca, the use of which underlines the sanctity of nature for the shamans who prepare and imbibe it. In his book *The Cosmic Serpent*, Jeremy Narby writes of his experiences with the Ashaninca people of the Upper Amazon, concluding that the ayahuasca shamans there work their magic through direct communication with the DNA that is the building block of all planetary life. Through ayahuasca, they go beyond their connection to the spirit of nature to arrive at the stuff from which nature and all things are made, merging with "the global network of DNA-based life."[3]

When he took part in ayahuasca ceremonies, Narby experienced visions of two gigantic boas that spoke to him without words. This fired his interest and he began to explore the consistency of such shamanic imagery. The first similarity he noticed was the common image of reptiles and snakes, often a "celestial serpent," that occurs in shamanic traditions the world over.

Mythologist Joseph Campbell also noted this, and wrote that, "Wherever nature is revered as self-moving, and so inherently divine, the serpent is revered as symbolic of its divine life."[4]

The similarities between DNA, the ayahuasca vine itself, and the snake imagery of the shamanic experience led Narby to suggest that shamans, through their ceremonies and journeys, are able to communicate directly with the information stored in DNA. He then began to study the characteristics of DNA and found that it emits electromagnetic waves corresponding to the narrow band of visible light. This weak light is equivalent to the intensity of a candle at a distance of 10 kilometers, but with a surprisingly high degree of coherence—comparable to a laser. It is fascinating to speculate that this may be the waveform of consciousness itself and that plants such as ayahuasca are the means of making it visible.

THE ROPE FROM THE MOON

Ayahuasca is the jungle medicine of the Upper Amazon. Made from the ayahuasca vine *(Banisteriopsis caapi)* and the leaf of the *chacruna* plant *(Psychotria viridis),* the two create a potent mixture that opens the person who drinks it to the experience of an energetic world underlying our own. Its very name suggests these properties, derived as it is from two Quechua words: *aya* meaning "spirit," "ancestor," or "dead person," and *huasca* meaning "vine" or "rope." Hence, the word *ayahuasca* translates as "the vine of the dead" or "the vine of souls," implying a means for communion with the spirit of the universe itself.

Both plants are collected from the rainforest in a ritual way that involves dieting and spiritual preparation, and it is said that the shaman can find plentiful sources of the vine by listening for the "drumbeat" or vibration that emanates from it. The mixture is prepared by cutting the vines to cookable lengths, scraping and cleaning them, pounding them to a pulp, and adding the chacruna leaves. The mixture is then boiled and reduced for about twelve hours until it becomes a thick, brown liquid. When drunk, this brew will produce a visionary experience lasting up to four hours.

There are mysteries surrounding how the shamans learned to combine these two plants to make up the brew, for without their combination each plant is more or less inert. In scientific terms, chacruna contains vision-inducing alkaloids and the vine is an inhibitor. It is the mixture of these that gives ayahuasca its unique properties. The main psychotropic ingredients in chacruna are tryptamines, which, if taken orally by themselves, would be immediately rendered inactive by the body's enzymes. The ayahuasca vine, however, contains monoamine oxidase (MAO) inhibitors in the form of harmine compounds, so when the two plants come together they complement each other. The resulting psychoactive compound has an identical chemical makeup to the organic tryptamines in our bodies. The mixture is therefore able to make its way easily into our brains where it bonds smoothly to our synaptic receptor sites, allowing a slow release of tryptamines into our bodies and a powerful visionary experience.

And yet the vine and the chacruna plant do not grow anywhere near each other. So how did the shamans know that they should combine them or where they would even find each one?

Simple. According to the Shipibo people, the plants themselves provided the answers. The following legend explains how.

> There was once a woman who was interested in plants and liked to pick their leaves. She would crush them in a pot and soak them in water overnight. Then she would bathe in them each morning before sunrise, knowing that the way to find out about plants and their effects is to be with them.
>
> One night she had a dream during which an old woman came to her and asked, "Why are you bathing in these leaves each day?" The younger woman recognized her visitor as the spirit of the leaves. "I am doing this because I want you to teach me," she answered. The old woman then said, "You must seek out my uncle. His name is Kamarampi. I will show you where to find him."*
>
> The young woman went to the uncle and he showed her how to pick the leaves of the chacruna, which was the bush she had taken leaves from to bathe in. He showed her where to find ayahuasca, which is the lover of chacruna, and how to prepare a marriage of them both. He told her to tell the people how to celebrate this marriage and how to use the brew.

Another legend, related by Amazonian shaman Javier Arevalo, is that the first shamans drank their ayahuasca without chacruna but the ayahuasca showed them that its lover, the leaves, was missing.

> The ayahuasca said that chacruna was the doctor that gives the vision and it needed to be added. My great-grandfather was among these first shamans and he responded, "But how shall we find this plant?" The ayahuasca answered, "You can find it by turning two corners." So they went into the jungle and turned

---

*The Shipibo word *kamarampi* actually means "purging medicine."

two corners and there was a woman who called to them. She led them to a bush which was chacruna.

Part of the mystery of these jungle legends is why the ayahuasca would be so keen for the shamans to find the chacruna and add its leaves to the mix. The answer to that is provided by another Shipibo tale, about the "Moon Man":

> Many generations ago our ancestors could climb the great rope into the realm where the spirits of the animals and the forest lived. Our ancestors and the spirits lived in both worlds at the same time. There was no separation.
>
> These ancestors could visit and talk with the plants and animals and they would share their knowledge of which plants to use for healing, which songs to sing to the animals we hunted. And we learned that we were at one with all life.
>
> Our ancestors lived in harmony and peace this way until one day, the Moon Man came to our people and severed the great rope to the spirit world and we lost our way into that place. It was a terrible loss to our people and there was much sadness.
>
> But then our ancestors remembered a way back to that world: the ayahuasca vine, which became the rope that we climb into the spirit realm.

In the Shipibo tradition, the Moon Man is associated with the analytical mind, and it is "rational thinking," therefore, that has severed our sacred connection to the cosmic mind of the spirit realm. This legend therefore speaks of humanity's need to reunite with the consciousness of the universe, using the rope (ayahuasca) to climb our way back to the oneness we once knew. Only then can we re-enchant the world through imagination and inspiration.

THE AYAHUASCA EXPERIENCE

Western science discovered the mechanism of MAO inhibitors (MAOI) in the 1950s. But by listening to the plants, the shamans knew of them centuries ago and, to quote Terence McKenna, "have brilliantly exploited these facts in their search for techniques to access the magical dimensions."[5] These shamans did not describe the MAOI process in terms of chemicals and synapses, of course, but in terms of spirit; and their knowledge was expressed in myth, not science. By hearing the wisdom of the plants and acting on the call of their spirits, however, the shamans knew the answers to some of the greatest mysteries of brain and universe hundreds or perhaps thousands of years before science even knew what questions to ask.

Harvard professor Richard Evans Schultes, widely regarded as the father of modern ethnobotany, remarked on the antiquity of the ayahuasca experience:

> There is a magic intoxicant in northwesternmost South America which . . . can free the soul from corporeal confinement, allowing it to wander free and return to the body at will.
>
> The soul, thus untrammeled, liberates its owner from the everyday life and introduces him to wondrous realms of what he considers reality and permits him to communicate with his ancestors. . . .
>
> The plants involved are truly plants of the gods, for their powers are laid to supernatural forces residing in their tissues, and they were the divine gifts to the earliest Indians on earth. The drink employed for prophecy, divination, sorcery, and medical purposes, is so deeply rooted in native mythology and philosophy that there can be no doubt of its great age as part of aboriginal life.[6]

One of the common experiences with ayahuasca, as Narby found, is the onset of visions with an initial series of fast-moving kaleidoscopic or geometric images, bright-colored masks, serpents, shape-shifting faces, and even cartoon characters. Although these can be entertaining and absorbing and information-rich in themselves, they are just the start of the process, the prologue to the real visionary encounter.

Our experience suggests that during this image-loaded stage, ayahuasca works to repattern the brain and alter consciousness so that those who drink it can enter into dialogue with its deeper spiritual intelligence. Once beyond the images, direct communication is possible with the ayahuasca *spirit*, and it is at this point that the real information is revealed. It is now, for example, that the ayahuasca will tell a shaman-healer what is wrong with his patient, what medicines to prescribe, or which spirit has caused the illness or malaise. It is now that the voice of ayahuasca also sings of the deeper intelligence that permeates the universe, and from which gifts of insight and self-awareness spill.

"Ayahuasca wants you to understand," says Javier, "and so it opens doors to different dimensions. Often the mind can be obstructed from accessing inner knowledge, but ayahuasca opens the mind to abstract things that can't be seen in the material world. If I hadn't had the [ayahuasca] experience, for example, I would not be able to believe that a tree can have its own world and spirit. But when you see these dimensions for yourself, little by little you begin to accept the mystery of it all."

These points are echoed by master shaman Guillermo Arevalo, one of the most powerful and respected elders of the ayahuasca tradition, who explains how the mixture can help and heal.

> Ayahuasca organizes the emotions and calms the nerves. Using it, many people who are depressed can discover their own solutions and recover their self-esteem. They discover their spiritual sides.
>
> People are out of balance from not knowing this side of themselves. Many think that being human entitles them to live as they like, but they are in fact not fully human if they believe that, because they have recognized only their physical side and ignored their spirit completely. It is very difficult for them to shake off the rational mentality that only believes in a physical world because our culture and education separates us from reality and tells us that progress is all about science and reason. This is even true of our religions, which are supposed to teach about God, but in fact they lead us away from Him.

For example, I first discovered ayahuasca when I went to Brazil to study nursing for seven years. I found that in Brazil, peasant people use plants more than drugs from the pharmacy and there were women who healed using prayers and *yaje* [another name for ayahuasca]. I was excited by this and when I returned to Peru I wanted to teach this to my people, but I found that certain religions were against the use of natural medicine and shamanism. I thought, 'This can't be! These plants are healers! Does God not want our people to be well?' And so I became *determined* to show my people the value of the old ways.

When we drink ayahuasca we evolve and gain power and lucidity. Then we can create actions that take form in the world, and change the future and the past too. If there is some trauma in the past, for example, it can come up through ayahuasca, but then it can be healed. That's what ayahuasca is for.

MAKING THE MEDICINE

In the shaman's world, all plants have a spirit that is, in essence, angelic. But they can also have human emotions like jealousy, vengefulness, and wrath. It is said that the spirit of ayahuasca is very jealous and that if the rules of its preparation are not respected its spirit may become resentful.

When it is being prepared, therefore, the shaman must watch over it constantly to prevent bad spirits from entering. The fire also needs tending regularly throughout the twelve hours of its preparation, and the shaman should follow a special diet during this time (see chapter 2). Sexual abstinence is also emphasized. But the most important thing, as with all magical work, is the focus and intention of the shaman, and at all times he must direct his healing energy into the brew.

This also means that not just anyone can be there to watch the brewing process, since the moral or spiritual quality of all present, as well as their adherence (or otherwise) to the diet, can all have an influence. The shaman's patient should not watch the preparation process either, nor should any woman who is menstruating, as this could leave misplaced energy in the medicine.

The question of menstruation is one that occurs in many different spiritual traditions, including Christian ones. Anthropologists find it hard to explain this taboo. One possible explanation is that the invisible messengers of the body, pheromones, can influence the visionary state. In many traditions it is said that a woman on her "moon time" takes away the vision of the shaman, and this is said to be true during ayahuasca ceremonies, where the changed hormonal balance and the subtle effects of aroma of a woman during her period can also alter the trance of those around her. In our experience, a woman who is menstruating is also more restless during ceremonies and sometimes this can disturb others.

Perhaps this is what our ancestors were prescribing against? Certainly there seems nothing to this taboo in terms of "sexual politics" and the Amazonian women self-regulate in this regard. They would never dream of entering a ceremonial space during their periods, for example, nor will female shamans prepare the brew at this time.

THE SONG OF THE SHAMAN

During the ritual preparation of ayahuasca (and certainly during its ceremonial ingestion), shamans often sing sacred songs, known as icaros. These may be magical chants or melodies that they whistle, sing, or whisper into the brew. They may also sing these directly into the energy field of a person who is to be healed during a ceremony.

An icaro can be regarded as an energetic force charged with positive or healing intent that the shaman stores inside his body and is able to transmit to another person or to the brew itself so that this positive energy is ingested when the mixture is drunk.

The songs are transmitted to the shaman by the spirit of the plant allies he has an affinity with, and the longer the relationship between shaman and plant, the more icaros he may learn and the more potent they will be. The power and knowledge of an *ayahuascero* (ayahuasca shaman) is therefore measured in part by the number of icaros he possesses. Javier, for example, has worked with many different plants for fifteen years and now knows the spirit songs of some 1,500 "jungle doctors," including the *icaro del tabaco* (the song of tobacco—one of the most sacred of Amazonian plants), the *icaro del ajo sacha,* and the *icaro del chiric sanango,* among the many others.

There are precise and specific icaros for many different purposes—to cure snakebites, for example, or to clarify the vision during ayahuasca ceremonies, to communicate with the spirit world, or even to win the love of a woman. *Huarmi icaros*—from the Quechua word *huarmi* (which loosely translates as "woman") are of this latter category. There are icaros called *icaros de la piedra,* which are taught to the shaman by *encantos* (special healing stones that offer spiritual protection), and icaros to the spirits of the elements, such as *icaro del viento,* which calls upon the spirit of the wind.

Other icaros, such as the *ayaruna*—from the Quechua words *aya* ("spirit" or "dead") and *runa* ("people") are sung to invoke the "spirit people"—the souls of dead shamans who live in the underwater world— so they may help during a healing or an ayahuasca ceremony.

Icaros can also be transmitted from a master shaman to his disciple but, as always, it is nature that is regarded as the greatest teacher; the most powerful songs are those learned directly from the plants themselves. To learn these songs, the shaman must fast or follow the diet for many weeks as she treks deep into the rainforest to find the appropriate plants and places of power where the magical music of nature can be heard.

A few verses from the *icaro madre naturaleza* (song of mother nature), which is chanted by Javier Arevalo, demonstrate this deep bond between the shaman and the natural world.

No me dejes no me dejes	*Don't leave me, don't leave me*
Madre mia naturaleza	*My mother nature*
No me dejes no me dejes	*Don't leave me, don't leave me*
Madre mia naturaleza	*My mother nature*
Por que vas i ti me dejares	*For if you will leave me*
Moriria o de las penas	*I would die of the pain*

Llantos y desesperaciones	*Tears of desperation*
Madre mia naturaleza	*Mother nature*
Si tu tienes el don de	*Yes, you have the gift of*
la Santa purificacion en ti manos	*Sacred purification in you hands*
Benditas madre naturaleza	*Blessed mother nature*
No me dejes no me dejes	*Don't leave me, don't leave me*
Madre mia naturaleza	*Mother nature*
No me dejes no me dejes	*Don't leave me, don't leave me*
Madre mia naturaleza	*Mother nature*
Por que vas i ti me dejares	*For if you will leave me*
Moriria o de las penas	*I would die of the pain*
Llantos y desesperaciones	*Tears of desperation*
El velo blanco que tu tienes	*The white veil you have*
Como cubre a esta criatura	*As it covers this child*
Limpia mi cuerpo y espirutu	*Clean my body and spirit*
Con el soplo o de tus labios	*With the breath of your lips*
Madre cita milagrosa	*Dearest, miraculous, mother*

THE CACTUS OF VISION

In the shamanic traditions of Northern Peru, meanwhile, it is not ayahuasca but the San Pedro cactus *(Trichocereus pachanoi),* or cactus of vision, that opens the doorway to expanded awareness and acts as mediator between man and the gods. San Pedro grows on the dry eastern slopes of the Andes, at altitudes between 2,000 and 3,000 meters above sea level, and commonly reaches six meters or more in height. It is also grown by local shamans in their herb gardens and has been used since ancient times, with a tradition in Peru that has been unbroken for at least three thousand years.

The earliest depiction of the San Pedro cactus is a carving dating from about 1300 B.C.E., showing a mythological being holding a San

Pedro. It comes from the Chavín culture (c. 1400–400 B.C.E.) and was found in a temple at Chavín de Huantar, in the northern highlands of Peru. The later Mochica culture (c. 500 C.E.) also depicted the cactus in its iconography, suggesting a continued use throughout this period.

Even in the present Christianized mythology of this area, there is a legend told that God hid the keys to Heaven in a secret place and that San Pedro (Saint Peter) used the magical powers of a cactus to find this place so the people of the world could share in paradise. The cactus was named after him out of respect for his Promethean intervention on behalf of mortal men.

As can be imagined, early European missionaries held native practices in considerable contempt and were very negative when reporting the use of San Pedro. One sixteenth-century conquistador described it as a plant by which the natives are able to "speak with the devil, who answers them in certain stones and in other things they venerate."[7]

As you might also imagine, a shaman's account of the cactus is in radical contrast to this. Juan Navarro, a maestro within the San Pedro tradition, explains its effects as follows:

> It first produces a dreamy state and then a great vision, a clearing of all the faculties, and a sense of tranquility. Then comes detachment, a sort of visual force inclusive of all the senses, including the sixth sense, the telepathic sense of transmitting oneself across time and matter . . . like a kind of removal of one's thought to a distant dimension.

Considered the "maestro of maestros," San Pedro enables the shaman to open a portal between the visible and the invisible world for his or her people. In fact, its Quechua name is *punku*, which means "doorway."

AN INTERVIEW WITH A SAN PEDRO MAESTRO

Juan Navarro was born in the highland Andean village of Somate, department of Piura. He is the descendant of a long line of healers working not only with San Pedro but with the magical powers of the sacred lakes known as Las Huaringas, which have been revered for their healing properties since the earliest Peruvian civilization.

At the age of eight, Juan made a pilgrimage to Las Huaringas and drank San Pedro for the first time. Now in his fifties, every month or so it is still necessary for him to return there to accumulate the energy he needs to protect and heal his people.

Healing sessions with San Pedro involve an intricate sequence of processes, including invocation, diagnosis, divination, and healing with natural power objects, called *artes*, which are kept, during the ceremony, in a complicated and precise array on the maestro's altar or *mesa*. (See a photograph of Juan Navarro's mesa in the color insert.) Artes may include shells, swords, magnets, quartz crystals, objects resembling sexual organs, rocks that spark when struck together, and stones retrieved from animals' stomachs that they had swallowed to aid digestion. The artes bring magical qualities to the ceremony where, under the visionary influence of San Pedro, their invisible powers may be seen and experienced.

The maestro's mesa, on which these artes sit, is a representation of the forces of nature and the cosmos. Through the mesa the shaman is able to work with and influence these forces to diagnose and heal disease.

What happens during a San Pedro ceremony?

JN: The power of San Pedro works in combination with tobacco [see below]. Also the sacred lakes of Las Huaringas are very important. This is where we go to find the most powerful healing herbs, which we use to energize our people.

We also use dominio* to give strength and protection from supernatural forces such as sorcery and negative thoughts. This dominio is also put into the seguros† we make for our patients. Dominio is introduced to the bottle through the breath. You keep these seguros in your home and your life will go well.

---

*Dominio is the linking of intent to the power of the plants.
†Seguros are amulet bottles filled with perfume, plants, and seeds gathered from Las Huaringas.

How does San Pedro help in the healings you do?

JN: *San Pedro helps the maestro to see what the problem is with his patient before any of this healing begins. The cactus is a powerful teacher plant. It has a certain mystery to it and the healer must also be compatible with it. It won't work for everybody, but the maestro has a special relationship with its spirit.*

When it is taken by a patient it circulates in his body and where it finds abnormality it enables the shaman to detect it. It lets him know the pain the patient feels and where in his body it is. So it is the link between patient and maestro.

It also purifies the blood of the person who drinks it and balances the nervous system so people lose their fears and are charged with positive energy.

In the ceremonies we've attended, a lot seems to happen. Can you explain the process to us?

JN: *Patients first take a* contrachisa. *This is a plant [actually, the outer skin of the San Pedro cactus] that causes them to purge* so they get rid of the spiritual toxins that are within their systems. This is a healing. It also cleans out the gut to make room for San Pedro so the visions will come.*

They also take a singado. *This is a liquid containing [aguardiente and macerated] tobacco which they snort through their nostrils. The tobacco leaf is left for two to three months in contact with honey, and when required for the singado it is macerated with aguardiente.*

How it functions depends on which nostril is used. When taken in the left nostril it will liberate the patient from negative energy, including psychosomatic ills, pains in the body, or the bad influences of other people. As he takes it in he must concentrate on the situation that is going badly or the person who is doing him harm. When taken through the right nostril it is for rehabilitating and energizing, so that all of that patient's projects will go well.

Afterwards he can spit the tobacco out or swallow it, it doesn't matter.

---

*To *purge* is, literally, to vomit. However, it also implies the removal from the body of spiritual and physical toxins, and is regarded in Peru as a highly beneficial form of healing.

The singado also has a relationship with the San Pedro in the body, and intensifies the visionary effects.

During the ceremony I also use a chungana [rattle] to invoke the spirits of the dead, whether of family or of great shamans, so they can help to heal the patient. The chunganas give me enchantment* and have a relaxing effect when the patient takes San Pedro.

What is the significance of the artes and of Las Huaringas?

JN: The artes that I use come from Las Huaringas, where a special energy is bestowed on everything, including the healing herbs that grow there and nowhere else.

If you bathe in the lakes it takes away your ills. You bathe with the intention of leaving everything negative behind. People also go there to leave the spirit of their enemies behind so they can't do any more harm.

After bathing, the maestro cleanses you with the artes, swords, bars, chontas,† and even huacos [the energetic power of the ancient sites themselves]. They flourish you—spraying you with agua florida‡ and herb macerations, and giving you things like honey, so your life will be sweet and flourish.

Not far from Las Huaringas is a place called Sondor, which has its own lakes. This is where evil magic is practiced by brujos [sorcerers] and where they do harm in a variety of ways. I know this because I am a healer and I must know how sorcery is done so I can defend myself and my patients.

As we said, a lot goes on in a healing! So, with all of this, just how important is San Pedro?

JN: What allows me to read a patient§ is the power of San Pedro and tobacco. Perceptions come to me through any one of my senses or through

---

Enchantment in this context means protection and positive energy.

†*Chontas* are bamboo staffs used as healing tools to lightly beat or stroke a patient in order to scrape negativity off him.

‡*Agua florida*, known as flourishing water in Peru, is a form of perfume containing healing spirits. In Haiti and some parts of the United States it is known as Florida Water.

§To *read* is to psychically diagnose a patient's problems by seeing her past, present, or future.

an awareness of what the patient is feeling; a weakness, a pain or whatever. Sometimes, for instance, a bad taste in my mouth may indicate that the patient has a bad liver.

Of course, I must also take the San Pedro and tobacco, to protect myself from the patient's negativity and illness, and because it brings vision.

HALLUCINOGENS IN THE WEST

The accounts of both ayahuasceros and San Pedro shamans and their descriptions of working with the plant doctors and allies that are their spiritual and healing partners reveal a very gentle, lyrical, and ensouled approach to the world. A very different picture begins to emerge from these accounts than the images we get from newspaper reports and television documentaries, which lead us to believe that hallucinogens of any description are bad and need to be banned.

The stories of the shamans tell of gifting their patients with power, of opening their minds to new possibilities and freedoms, to a life of connection and a sense of the divine. By contrast, our modern view seems based in fear and distaste for the very possibilities of such a connection. What can account for the discrepancy between the two?

We are not the first writers to remark on the political dimension to all of this. For example, the comedian Bill Hicks writes: "Drugs that grow naturally upon this planet, drugs that open your eyes up, to make you realize how you're being fucked every day of your life. Those are against the law. Wow! Coincidence? I don't think so. . . . It's not a war on drugs, it's a war on personal freedom."[8] The governments of the Western world want to ban hallucinogens (and, indeed, have already done so) and prosecute those who use them. But what is really being controlled? Not the drugs, but our freedoms and our minds. The hidden message behind such prohibitions is, "We will not allow you to expand your consciousness beyond the norm of socially prescribed reality or to see possibilities in the world that do not come from us."

This is an insidious message. Digging further, what it really says is: "We will not allow you to be fully human, to use your mind, to experience reality for what it is. We withhold the sacred from you." But our minds are our own! Surely we have every right to explore them as we wish?

Sacred hallucinogens, furthermore, are not "drugs" but plants, and surely there is a fault in the logic that says that plants can be illegal substances, when they grow naturally, broadly, and bountifully from the Earth and in every country of the world. Bill Hicks again: "Making cannabis illegal is like saying God made a mistake." The land gives freely to all. Where is the logic in saying it should be owned and controlled by the few?

Argue or reason as we might, we in the West are facing a situation of increasing repression of our freedom to experience the sacred, as plants like ayahuasca and San Pedro help us to do. We are losing our connection to the spirit of the world because of the power of government to control our access to it. This repression is not for our safety—to stop us from "getting high" and jumping off buildings—but to keep us locked in a materialistic mindset. Any transformational experience that shows us the interconnectedness of all things, and our part in the planetary consciousness, runs contrary to this objective. But it is obvious how self-defeating this objective ultimately is when we look around at where such materialism has got us.

Everything we do to our planet in the name of "progress," we also do to ourselves. As we lose the power to dream, so our dreams die and we create a world based in fear and conflict—the very things for which the West has become known, and a far cry from the gentle world of the shaman.

EXPERIENCING THE SACRED

It may be impossible for you to experience sacred hallucinogens and they may already be illegal in your country, as ayahuasca now is in the United States. So we need to look at other creative ways to enhance our neurochemistry if we wish to experience the sacred through an expansion of consciousness.

The key here is that many of the compounds and neurotransmitters that are present in sacred plants also occur naturally within the human brain and can be enhanced through specific practices. Dr. John Lilly found, for example, during his work with flotation tanks and altered states of consciousness, that when the mind is deprived of external stimuli, it opens to a number of unusual sensations and spiritual effects. These include waking visions, lucid dreams, and even a kind of out-of-body travel.[9]

Reducing outside stimulation is therefore one key to the ecstatic experience. And so, for the first exercise below, it is useful to lie down and relax in a quiet, darkened room, having set aside time (about an hour) in which you will not be disturbed. Cover your eyes and ears so that you fully block out the world.

Dreaming the Great Spirit

The shaman's intention is a calling, a powerful summons to the universal field of energy or consciousness of which we are an intrinsic part, and which generates a movement of the spirit toward the expanded mind. Begin, then, by setting your intention to meet with the spirit of nature.

In both the Amazon and Haiti, this spirit is believed to be the forest itself. The trees and the leaves—the entire ecosystem, in fact—are the visible faces of this spirit.

In Haiti, this spirit has a personality and is an aspect of God (known as a Lwa) that can physically possess people during Vodou ceremonies and, through them, heal the others who are present. His name is Gran Bwa (which translates from the Kreyol language as "Big Wood")—he is lord of the great forest and presides over the deepest mysteries of healing and of initiation into the spirit world.

Each of the Lwa has a particular symbol or pictorial representation of his or herself, known as a *vever*, which is drawn on the earth in cornmeal during a ritual to call this spirit. Gran Bwa's is a drawing of a leaf that also has human features to it—a face, arms, and legs—further signifying his (and, indeed, humanity's) connection to nature.*

One way to conceive of the spirit of nature, then, is to imagine yourself on a vision quest through a vast forest, which is alive and breathing and holding you as you move through the trees. Without veto, allow any images, thoughts, or feelings to simply drift through your mind as you make this quest. These are the whispers of spirit from beyond the rational world.

Eventually you will arrive at one special tree that holds a particular fascination for you. This is Gran Bwa, the cosmic doorway to the whole of nature. The classic four questions of the vision quest are: *Who am I? Why am I here? Where am I going? And who will help me in this?* Ask these questions of this spirit, and be open to the answers you receive.

When you have the information you need, bring yourself back to ordinary awareness. Then physically go outside into nature and see how it looks to you now. Repeat this exercise once a week when you diet and see what else you can learn.

A journalist from the *London Observer* newspaper attended a retreat run by Ross in 2005, where she undertook an exercise similar to this and was amazed at what she saw in nature once she looked at it for the first time in a quiet and nonjudgmental way and understood it as a living force. "The sight of nature in all its majesty was overwhelming," she wrote in her newspaper. "I could see everything. From the tiniest hair on the outside of a leaf to the iridescent sequins on the inside of a petal and the minuscule contours on the body of a dragonfly.

"But more than that, I felt all of this too. It was like I had developed another layer of perception. Beyond what I saw, I could sense. Even now, a week on, I can bring this feeling back . . . "[10]

---

*There is more information on Gran Bwa and Haitian healing in *Vodou Shaman* by Ross Heaven (Destiny Books, 2003).

THE ICARO: A SONG OF YOUR SOUL

Just as there is a vever for each Lwa in Haiti, there is at least one song (and usually several) for every spirit. These songs are not so much created as "discovered" by the shaman, who enters into trance communion and allows the spirit itself to sing its song through him. Here is one of the songs for Gran Bwa:

Se nan bwa, fey nan bwa ye,	*It's in the woods, the leaves are,*
Se nan bwa, fey nan bwa ye,	*It's in the woods, the leaves are,*
Se mwen menm, Gran Bwa	*It is I, Gran Bwa*
M pap montre moun kay mwen,	*[But] I won't show people my house,*
Si m pral montre moun kay mwen,	*If I show people my house,*
Yap di se nan bwa m rete	*They will say I live in the woods*

The meaning of the last three lines, and especially the final one, is that while the deep woods are, indeed, the home of Gran Bwa, that is not the *only* place he can be found, for he is the spirit of nature itself and a part of everyone and all there is.

The spirit songs and chantes of Haiti are similar to the icaros of the ayahuasca shaman in that the spirits themselves have taught these songs of power. By singing them aloud (or allowing himself to be sung) the shaman brings the vibration of the healing universe and the powers of nature into his body, opening himself to a deeper level of awareness through this union of spirit and matter. When such songs are sung in Haiti, they are calls to the spirit, who may answer by possessing the shaman so he can heal the community. When such songs are sung in the Amazon, their vibration may be blown from the shaman into the body of his patient, healing him directly through the powers of nature.

🐚 Discovering Your Ally's Song

Another way for you to more deeply experience the sacred, then, is to know the song of your plant ally. How we discover it is simplicity itself. We ask.

Lie down in the journeying posture you are familiar with, holding your plant ally against your heart, and then allow your consciousness to drift back to that great tree in the forest that called you in the last exercise. See yourself standing before this ancient teacher, your plant ally in hand, and ask the tree itself for the song.

When you feel in your body that there is a tune, a vibration, or words that are aching to be sung, simply allow them to flow from you. There is no need to search for words (indeed, some songs do not have words at all; they are wordless chants or even metalanguage, which sounds like "speaking in tongues").

Bring yourself back to normal awareness when you have your song, singing it as you do so, and then spend some time with your plant ally, singing its song to it. Listen for any words that come back to you from the plant itself, which you will hear and understand in your heart and in your mind.

THE SEGURO:
A FRIEND WHO WILL LISTEN

According to San Pedro maestro, Juan Navarro, a *seguro* is a friend or ally, someone you can turn to for advice and information, who will listen and share your problems. Less poetically, a seguro is a clear glass bottle that contains perfumes, sacred water, and of course a selection of plants chosen for their specific healing and spiritual qualities. These bottles are kept on an altar, in sacred space, and are regarded as objects of great power. Whenever the person who has a seguro requires help with any practical or spiritual problem, she will take it from the altar and sit with it against her heart, speaking with it as if to a friend. The seguro will absorb and transform the energy of the problem, but more importantly, if she listens carefully, the person who seeks its advice will hear the answers she needs from the spirit of the plants themselves.

A seguro can help you maintain and deepen your link to the sacred because, of course, it also contains your plant ally. If there are other plants you have journeyed to or would like to learn from, you can add these to the seguro as well and, now that you know the language of your ally, this plant spirit will communicate your desire to the other plants, which will also offer their healing and support. You therefore gain access to the natural world and its powers more widely.

Creating a Seguro

To create a seguro, you will need a glass bottle, approximately five inches high, that can be sealed. Fill this one-third full with perfume of your choice and top up with water. In Juan Navarro's seguros, this is water from the sacred lakes of Las Huaringas, but you can use mineral water or rainwater (as pure as possible).

Once you have this base, meditate for a while on the qualities you would like in your life and which plants might bring you these things. Be informed in this by your work with the doctrine of signatures—heather for luck, honesty for truth, goldenrod for wealth, and so on.

Add these plants to your bottle, arranging them as attractively as possible. (Some seguros are so beautiful they are works of art in themselves.) Then place your main plant ally in the bottle so it can act as mediator for all the others. Before you seal the bottle, blow your dominio (intention) into it three times, and then put on the lid.

Place the bottle on your altar and reflect on its qualities often. Whenever you are in need of advice, sit with your seguro and speak with it. Then notice how things change for you.

Maintaining the Sacred Communion: Eating for a Healthy Neural Base

Shamans who work with sacred plants undertake a diet that enhances their connection to nature and their allies (see chapter 2). Diet—in its widest sense—is important for us, too, and we need to look at the plants, herbs, and amino acids that will increase the level of neurotransmitters in our brains, as these will help to develop our skills of shamanic dreaming.

Brain cells require a supply of vitamins, fatty acids, and minerals to function effectively and for us to be healthy. Low nutrient intake is a growing problem in the West as a result of our fast-food lifestyles and "quick and junky" eating habits. As a consequence, more and more people are suffering symptoms such as depression and anxiety.

When our brain cells die off through nutrient depletion, they may not be replaced, so our first step must be to protect our existing cells from degeneration. The plant medicine compounds that can help with this are known as adaptogens and antioxidants. You are probably familiar with the names of some of these adaptogens: panax ginseng, for example. These plants help the body to resist stress and prevent damage to the cells. Another compound that exists in the human body and needs to be strengthened is the enzyme Co-Q10. This is available in many supplements and in foods such as papaya, which is loaded with powerful antioxidants and vitamins.

The B vitamin group is important for improving cognitive performance and enhancing learning and memory. A number of studies have shown B6 and B12 deficiencies in elderly people, resulting in cognitive impairment such as memory loss. Vitamin B6 supports the synthesis of neurotransmitters such as serotonin and dopamine; it is found naturally in a wide range of cereals and foods such as beans, bananas, oatmeal, peanuts, and chicken. Vitamin B12 is essential to maintaining healthy nerve cells and is found mainly in foods such as oysters, clams, liver, trout, salmon, eggs, and dairy foods. Vegetarians and vegans can take this vitamin in supplement form.

Other plants raise serotonin levels in our brains to give us a more powerful capacity for dreaming, as well as feelings of well-being. Examples of serotonin-boosting plants include black cohosh and bananas. Serotonin interaction within the brain is one of the key factors in the hallucinogenic experience.

Omega-3 fatty acids are particularly important for the health of the brain, and these can be found in fish and flaxseed oil. The importance of these fatty acids can be seen in cases where they are deficient. In newborn children, for example, omega-3 deficiency is associated with delayed visual and cognitive abilities, and low levels have also been found in people with Alzheimer's.

There are foods you should avoid as well—foods that work against our neural well-being and cognitive abilities. Key among these are processed sugars and carbohydrates; but any form of processed food should, as much as possible, be avoided.

As a Westerner wanting to develop your capacity for shamanic dreaming, your basic diet should therefore include lots of fresh fruit, vegetables, rice, fish, and chicken; and should exclude fried, fatty, over-spiced, or processed foods. This would be a diet, in fact, very similar to that of an Amazonian or Haitian shaman. Practice this diet to the best of your ability for the length of time that you are developing your relationship with your plant ally—and beyond.

Art and Action

In addition to eating properly, there are certain actions we can take to increase our dreaming abilities. Art and craft work is a good example. We don't need to be trained artists to work with color and form. In fact, training may be counterproductive in this case, since training in art to some extent involves the use of the rational mind instead of the dreaming senses, so that we *study* form instead of *responding* to it—and this is exactly what we are aiming to avoid.

To use art shamanically, your intention instead is to *play*, to become a child once again and to experience a child's sense of awe and wonder at the world. As Goethe said, "To know how cherries and strawberries taste, ask children and birds."

Draw, paint, sculpt with clay, carve with wood or stone, weave, embroider, write poetry, garden . . . do whatever inspires you and gives reign to your creative expression. During the artistic process, notice, as a child would, the forms in front of you, but also the spaces and shapes between the forms. Working in this way gives the nonrational mind a wonderful workout.

Other practices to boost your dreaming abilities are meditation, walking in nature, and simply being still and remembering that we are on this Earth to enjoy ourselves!

Rejoice in your body. Dance, sing, drum, play your flute. Dance, movement, and music are all keys to personal freedom, and at the same time an intrinsic part of expanding our consciousness and reconnecting to our amazement at the "spirituality of all things," as Walt Whitman described it.

AYAHUASCA ANALOGUES

In America and Europe, it is possible to create an analogue for ayahuasca by using plants that contain the same alkaloids, if such plants are not illegal. Seeds of syrian rue can be used as a substitute for the vine, for example, and mimosa or acacia, which are both rich in DMT (tryptamines), can be used instead of chacruna.

The Web site www.erowid.org is a good source of information for these plants as well as for other sacred hallucinogens.

4

HEALING THE SOUL

The soul cannot thrive in the absence of a garden. If you don't want paradise, you are not human; and if you are not human, you don't have a soul.

THOMAS MOORE

The world over, since the dawn of human experience, we have used plants to replenish lost vital energy and to remove negative spiritual influences—the two principle causes of illness for traditional societies. Neither of these causes have much, if anything, to do with conventional medicine, of course. Instead, ill health or "dis-ease" is seen as one sign of an inner state of disequilibrium or disturbance, which, from a shamanic perspective, will usually arise from two forms of energetic imbalance:

Spirit Intrusion—where forces external to a person find their way into that individual's energy system, either absorbed and assimilated from people around him or her, or, in some cases, introduced deliberately into it. These forces can result in physical, emotional, mental, and spiritual discomfort.

Soul Loss—where traumatic events, such as accidents, the loss of a family member, natural disasters, or the experience of abuse result in the severe loss of a person's life force, which, again, creates imbalance and illness.

Both of these conditions can leave a person disinterested in life, with no desire or ability to engage in it, and may eventually lead to physical illness, as well as more psychospiritual problems, because the body as well as the life force is weakened. In this chapter, we look at some of the causes and symptoms of soul loss and spirit intrusion, as well as healing actions and preventive measures we can all take, with the support of our allies in nature.

SPIRIT INTRUSION

Behind the notion of spirit intrusion as a cause of illness is the understanding that human beings, like nature as a whole, are composed of energy, and this energy can become weakened in certain circumstances. When that happens, we become vulnerable, because our innate ability to resist negative or harmful influences is reduced. We then become increasingly susceptible to viruses, infections, accidents, emotional wounds, and mental distress, as well as *saladera* (bad luck), as our life force is weakened in this way.

These influences (some shamanic traditions call them *essences*) are around us every day in the form of energy emanations from other people and places, and the subtleties of mood carried on the atmosphere that surrounds us. All of these may affect our outlook and serenity; but under normal circumstances, when our life force is strong and in balance, they do not pose a major problem. When we are not feeling powerful, however, a field of energy greater than our own can easily overwhelm us.

Key among the reasons we lose personal power, and thereby become vulnerable to potentially damaging influences, is allowing ourselves to become disconnected from nature. The world of nature is our first source of power and will normally keep us safe by "recharging our batteries" so we feel strong.

"When a man or woman is out of balance with the natural world, the energies of other people are easily absorbed by their system and can start to attack them," says Antoine Duvalier, a healer in Port-au-Prince, Haiti. "Because this energy is not natural to him, he does not benefit from a boost of power; it clogs his spirit instead and feeds from it, leaving him drained of power. In some cases, rivalry, jealousy, and the struggle for supremacy mean that a *mo* [bad energy in the form of a dead spirit] is deliberately sent after a person, with the intention of harming him."

The latter situation—a deliberate energetic or psychic attack—sounds malicious and extreme, but is far more common than we might think. Malidoma Somé writes of it occurring among the Dagara of Burkina Faso, where magical darts, visible only to shamans, are left on the ground in the path of an enemy so they embed themselves in her spirit when she walks over them.[1] Martin Prechtel writes of a similar practice among the people of Mexico and Guatemala.[2] And even in Western culture, we know such things are possible. We talk of people "looking daggers at us," and observe that "if looks could kill" we would be in trouble from those who mean us harm; these expressions are an unconscious acknowledgment of a spiritual fact: that looks and words—energy sent outward in symbolic form—have power. We know from personal experience that in arguments, for example, finger-pointing and unkind words can be physically felt. Who hasn't been wounded by the words of another person? We feel them in our guts.

"In some ways," says Michael Harner, "the concept of power intrusion is not too different from the Western medical concept of infection," and, in fact, instances "occur most frequently in urban areas . . . people, without knowing it, possess the potentiality for harming others with eruptions of their personal power when they enter a state of emotional disequilibrium such as anger. When we speak of someone 'radiating hostility', it is almost a latent expression of the shamanic view."[3]

Harner's words can also be seen as a statement, not on power, but on the *powerlessness* that people feel in modern society, for if we truly had equality, equilibrium, and inner strength, and our souls were nourished by our world, there would be no need for the jealousy and frustration that leads to eruptions of anger. Spirit intrusions in the West are often energy projections (conscious or unconscious) from the disenfranchised urban man or woman, which are sent to take power from another in the hope of gaining a sense of order in a world that seems out of control.

There is always something to envy in a world based on separation and the desire for material wealth, and where consumerism teaches people to want more and remain dissatisfied with what they have in their lives.

This competition, jealousy, or neediness is also at the root of intrusive illness in traditional societies. In many countries, envy is considered to be the cause of spirit intrusion. This may manifest as *mal d'ojo* (giving someone the evil eye), which is an infection of power in itself. But it may be more insidious: the sending of black magic toward someone, which causes them to become weak and lose whatever they have that has prompted the jealousy, even if the envious person gains nothing from it except the satisfaction of seeing his neighbor suffer.

To establish the cause of such a spirit-driven illness and find out who has sent it, an Andean shaman may conduct a divination with coca leaves or tobacco (see chapter 1). To remove it, she then needs to *see* it. *Seeing* is a practice of looking into the soul of a patient by becoming aware of the subtle emanations of energy that are given off by his physical body, much as you practiced it in the plant-gazing exercise in chapter 1. By doing this, the shaman becomes aware of energy that is out of resonance with the patient's own, recognizing it because its appearance will be in some way unpleasant for her to look at. Sandra Ingerman puts it this way: to the shaman, "All illness has a spiritual identity. It will look like a fanged reptile, or an insect, or some dark, sludgy material. The illness will show itself in a form that is repulsive."[4]

This does not mean, of course, that there is *actually* a reptile or insect within the patient; the shaman's seeing, or perception of it in this way, is a diagnostic device that allows her to recognize the intrusive energy so she will not miss or mistake it. Having said that, it is interesting how such energies do take on consistent forms. Ingerman says, for example, that: "Some of the modern work being done with imagery and healing correlates with what shamans have always seen with illness. For example, when patients with cancer draw what their illness looks like, they often draw fanged reptiles and insects. People see on their own what classical shamans have always seen."[5]

To test the truth of this, you have only to think back to a recent illness or localized pain of your own and ask yourself, "If I were to draw this, what would it look like?" People with skin irritations, for example, often describe their discomfort as "like an army of ants crawling over me," and depict it as such, while those with stomach problems say their pain is like "a snake writhing in my guts." It seems natural to use words and pictures like these to make our illness clear to others.

Once the shaman sees the intrusion, there are various ways he might deal with it in order to remove it. One of these is negotiation.

Eduardo Caldero, a curandero from Cusco, believes that spirit intrusions are not "evil" entities, but energies that are in themselves ambivalent to human beings. The person who sent them may have evil intent, but the energy he directs is often innocent, unaware that it is harming anyone through its presence, and simply trying to survive and live its own life.

> These spirits are like a homeless person finding an empty house—the empty house being a person without power. The homeless spirit wants to stay there, of course, and make a home and be happy. It doesn't intend its home any ill and has no desire to harm it—who would?! If you try to throw it back on the streets though, it will fight to stay where it is. That is only natural.

Caldero gets around this by offering the spirit a new home on the condition that it leaves the body of his patient.

> This is taken very seriously. It is a contract we [the shaman and the intrusion] make between us and there is a ceremony that goes with it.

CLEANSING THE SOUL WITH PLANTS

This ceremony takes the form of a *limpia*—a cleansing—where the patient sits in a wooden chair below which is a bowl of smoking copal incense. This will purify the patient's body and is relaxing to the intrusion, which is made drowsy by the smoke. As the limpia takes place, don Eduardo circles the patient, chanting, blowing tobacco smoke over her, and stroking her body with flowers (marigolds are often used). The tobacco smoke eases the passage of the intrusion, which is then caught by and rehoused in the flowers.

(Again and again we find that tobacco is used for cleansing negative energy. In the Amazon, tobacco—which is very different from that in your pack of cigarettes—is regarded as one of the most sacred plants. Wild tobacco is rolled up and lit and the smoke blown as a conduit of intention, the curandero engaging with the spirit of the plant as a focus for his energy, in a similar way to the North American Indian custom of using sage and sweetgrass in bundles or smudge sticks. It is never a mechanical action. In fact, when using tobacco, or any plant, to cleanse a space or a person, it is important to be in a state of 100 percent attention so you can see where the smoke goes and where it doesn't go. If the smoke is drawn into the body, this is often regarded as a healthy sign, but if it is pushed away from the body, the shaman knows that there are negative or blocked energies present to be worked with.)

Sometimes an offerenda is also made in thanks for the healing—or in thanks to the intrusion for leaving—in which case a gift of some kind may be tied up with the flowers. The whole bundle is then taken into nature and buried, so the spirit will not be disturbed and others won't be infected by it. Coastal shamans may take the flowers to the sea instead and cast them into the waves so the tide takes them away from the shore.

In the Amazon rainforest, the limpia uses not flowers but the leaves of the chacapa bush. (See a photo of chacapa leave on page 2 of the color insert.) These are approximately nine inches long and, when dried, are tied together to make a medicine tool that is used as a rattle during ceremonies. In a healing, the chacapa is rubbed and rattled over or near the patient's body to capture or brush out the spirit intrusion. Once he has it in his chacapa, the shaman then blows through the leaves to disperse the intrusion into the rainforest where the spirits of the plants absorb and discharge its energy.

Another way of dealing with intrusions is through the use of cleansing leaf baths, a method practiced in Haiti as much as in Peru (also see chapter 6). Loulou Prince explains:

> There are specific leaves, strong-smelling leaves, which help people who are under spiritual attack. I mix these leaves with rum and sea water to make a bath for the person, then I bathe her and I pray to the leaves to bless her. I sing songs for the spirits and the ancestors as well, and ask them to come help this person.
>
> The rest of the bath that is left over, I put in a green calabash bowl or a bottle, and before the person goes to sleep at night, I have her rub her arms and legs with it. When that is done, no curse can work on that person and the evil is removed.

How this "evil" comes to infect a patient in the first place also has to do with jealousy. For example, Loulou was asked to perform a healing for a young child brought to him by a woman who had four children, two of whom had already died through the actions of spirits that came to her house at night to suck the life force from them. The woman was a market trader who had made a little money (a rare commodity in Haiti). Her neighbor was jealous and had sent spirits to kill her children.

> I bathed the child to break the bad magic. Then I gave him leaves to make his blood bitter, so it would taste and smell bad to the spirits, and they would go away. After that, the child got better; he got fat and he grew. That boy is a young man now.

Intrusive spirits like these are believed, in Haiti, to make their home in the blood, which is why Loulou uses herbs to make the blood taste bitter and the body smell "strong." This makes the host less appealing to the intrusion, which then finds its way from the body.

Fire baths are often used in these treatments as well, where kleren becomes the base for an herbal mix, which is set on fire and rubbed over the skin. The alcohol burns quickly and doesn't hurt the patient, but it destroys the intrusion as it makes its way out of the body.

SIN EATING AND SPIRIT EXTRACTION

Dr. Stanley Krippner, professor of psychology at Saybrook Institute, and author of *The Realms of Healing*,[6] concludes from his study of traditional healing that the power of our thoughts alone—whether positive or negative—has a profound effect on our health. When we accept the psychic emanations of others, pick up on their negativity and—crucially—when we allow their negativity to be absorbed within us so we find ourselves in agreement with our enemies, we open ourselves to illness.

This, too, is the basic philosophy of sin eating. In this old Celtic tradition, a sin is viewed as a weight or blemish on the soul that will keep it suffering while alive and Earthbound when the sinner dies. Sin is a powerful force toward illness while we are alive, but it is *our knowledge* that we have done wrong that really creates the weakness in our souls. The shame and guilt we carry *is* the spirit intrusion.

This notion of sin as intrusion is also known to Tuvan shamans, who refer to sin as *buk*—an illness that arises through our malicious actions toward another person or life form. The Tuvan shaman Christina Harle-Salvennenon[7] gives the example of two young boys, patients of hers, who got carried away one day while they were playing and castrated a dog. When they came to their senses and realized what they had done, the boys ran home in shock. Both of them immediately became ill, one symptom of which was inflammation of their testicles. Recognizing the illness as buk, Christina demanded that the children tell her what they had done to cause its onset. The children, however, were overcome with guilt at their actions and refused to confess. Had they done so, it would have relieved the traumatic pressure in their bodies and given the shaman a direction for healing, but they simply could not. Both children died.

Spirit extraction (the removal of intrusions) was sometimes performed by the Celtic sin eater by stinging the patient's body with nettles, paying particular attention to the "corners and angles"—backs of the knees, elbows, back of the neck, and belly—where intrusions tend to congeal. The nettle stings would heat the skin and draw the intrusion to the surface, in a similar way to the fire baths of Haiti. It could then be washed off in a cold bath containing soothing and cooling herbs such as chamomile, lavender, rose water, and mint.

Once this was done, the patient would also be reminded of the need to make reparation to the person he or she had sinned against, or else the patient's guilt—and so the intrusion—might well return. One simple tradition that has survived as a way of making amends for minor sins, of course, is to send a bunch of flowers.*

HAITIAN EXTRACTION MEDICINE

Pakets kongo are one Haitian method of extraction. To look at, these are similar to mojo bags (see chapter 1), but more highly decorated and containing densely packed magical herbs that are specially blended to make them appealing to the spirits.

The herbal contents are a closely kept secret (though they do vary according to who creates the paket), but it is quite possible to make your own using local herbs once you understand the plant spirit principles behind this medicine tool. The secret is that each paket contains, in equal measure, plants that are (1) healing to the patient, and (2) attractive to the spirit intrusion. As the paket is used on the patient, therefore, intrusive energies are drawn into it, while the patient is soothed by the healing plants.

In Western cultures, soothing and purifying herbs for the patient might include the following:

---

*Sin-eating philosophy, again, is in many ways consistent with the Haitian experience. Maya Deren writes, for example, that therapeutic actions may be "executed by the priest but must be carried out, in major portion, by the patient himself under guidance of the priest. The patient must himself straighten out his difficulties with the loa. [*Loa* is another spelling of the Haitian word *Lwa*: spirits] . . . In other words, *the patient treats himself,* and this is another boost to his morale. Almost inevitably, no matter how ill the person is, he must take part in the rituals relating to his treatment."[8]

- Chamomile
- Marigold petals
- Bay
- Cedar
- Hyssop
- Hops
- Lavender

- Balm
- Parsley
- Rosemary
- Holy thistle
- Valerian
- Vervain

The herbs that are attractive to spirits tend to be hot and spicy:

- Cinnamon
- Clove
- Coriander
- Damiana
- Cactus
- Chili pepper
- Fig

- Garlic
- Ginger
- Hawthorne
- Mustard
- Paprika
- Peppercorns
- Wormwood

A selection of these is made (it is not necessary to use them all, which is what accounts for the variance in paket mixes), and the herbs are mixed with rum and Florida Water and allowed to dry. A roughly orange-sized ball of the mix is then placed in muslin and tied to make a bag. Around this a second skin of silk or satin is created, using colors and decorations such as sequins, feathers, lace, shells, and beads to reflect the qualities and preferences of specific protective spirits, and to energize the package with the *ashe* (power) of these spirits. The colors used correspond to the archetypal qualities we tend to be familiar with (see chapter 1, on mojo bags), so that red is for Ogoun, for example—the warrior spirit whose particular gift is power—while green is for Gran Bwa, the healing spirit of nature.

In Haiti, pakets are always made at night, usually beneath a full moon, and are dedicated to Simbi Makaya, the spirit of transformational magic. They are used in healing by passing the paket over the body of the patient, slowly, from head to toe, seven times. With each pass, the paket grows heavier as the herbs absorb intrusive energies while transferring healing to the patient. At the end of this, the paket is placed on the altar, which is a gateway between worlds and will discharge the intrusive energy back to the spiritual universe.

Used in this way, the power of each paket will last for seven years and is equally effective when used in a modern urban setting as when it is used in rural Haiti. Deirdre, a therapist from London who made a healing paket during one of our workshops later wrote, for example, that: "I had an *amazing* session today using the paket! It has powerful and different qualities —clearing and soothing. My clients also comment on it and remark about its powers."

SOUL LOSS: THE FRACTURED SELF

Soul loss is in many ways the exact opposite of spirit intrusion. In the latter, some alien force is injected into our spirit-selves; in the former, it is a loss of spirit, a depletion of energy, that causes the illness.

In the West, we know something of soul loss, though we use different terms for it. We talk, for example, of "psychological dissociation" and "stress-related syndromes," which arise as a result of traumatic, abusive, and hurtful experiences and may have a number of physical, mental, and emotional symptoms associated with them. In shamanic cultures, the same symptoms are diagnosed as a fracturing of the soul, or simply as soul loss, where an individual's spirit, faced with pain, has split into many parts, some of which have taken refuge in the otherworld, away from the harshness of everyday life.

In some ways, then, soul loss can be seen as a positive act of psychological and emotional self-protection in the face of great pain and distress. When we are suffering, of course we do not want to be present, and so our souls and our psyches have found a way for us to take refuge in a world away from our trials.

When the life force remains lost, however, even when the threat to the self is over, this can be equally debilitating. The patient may find himself disconnected from life and out of balance, so that his emotions, thoughts, memories, bodily reactions, and spiritual ambitions are out of alignment with his own true nature and that of the wider world and spiritual universe. This is when problems really begin and when the healing intervention of the shaman is most needed.

Soul retrieval—the recapitulation of life force—is one of the shaman's most effective healing practices for the restoration of this energy. To understand how it works, we need to remind ourselves of the shaman's perspective on reality: that it is multidimensional and operates beyond the constraints of time and space. From this perspective, anything that has ever happened to anybody, anywhere, is still happening somewhere. That is, even if a traumatic event occurred ten or twenty years ago, for the person who suffered it, it is still happening *now* because, if undealt with, it still influences his life and comes out in his behavior, which is an adaptation to the effects of that twenty-year-old event and the soul loss that accompanied it.

For the shaman, there is no "past," only one vast, awesome, ever-moving now. In his healing approach, he will therefore journey—outside of time and space—to the place where that energetic event is still happening for that individual and find and bring back the life force that is held there. Only when this has been done can the healing of the event and its consequences really begin.

The concept of soul loss, and the ceremonial retrieval of souls in this way, is found in many shamanic cultures. In the Tibetan Bon tradition, for example, one of the most important practices performed by shamans of the Sichen path is *lalu* (literally, "redeeming" or "buying back" the soul), and *chilu*, ("re-membering" the life energy). But, of course, this practice is far from restricted to Tibet. Joseph Campbell writes in *The Masks of God*, for example, that

> Sickness, according to shamanic theory, can be caused . . . by the departure of the soul from the body and its imprisonment in one of the spirit regions: above, below, or beyond the rim of the world.
>
> The Shaman's clairvoyant vision must discover its lurking place. Then riding "on the sound of his drum," he must sail away on the wings of trance to whatever spiritual realm

may harbour the soul in question, and work swiftly his deed of rescue.[9]

Although the terminology is different, the idea of soul loss is also well known to psychology. In his memoirs, Jung recounts a fantasy in which his soul flew away from him; that is, his libido withdrew into his unconscious to carry on a secret life there. In this example, the libido is the life force, and the unconscious is the "land of the dead," where unhappy souls go to hide.[10]

One of the few differences between shamanic healing and Western psychological systems, in fact, is that soul retrieval focuses on the return and integration of lost life force, rather than the original trauma itself, as psychotherapy tends to do. But soul retrieval and therapy do work very well together. The best combination seems to be: first, the recapitulation of life force, and then a therapeutic approach to support the patient through the process of understanding this newfound energy, the layers of meaning behind it, and the feelings and emotions that return with restored life force and are sometimes raw and uncomfortable.

This release and reintegration of previously subdued emotions can be fundamental to the whole healing process. Afterward, the patient can move forward in life without being anchored to the past and can live with new creativity and productivity.

LISTENING TO THE SOUL: HOW SOUL LOSS FEELS

In the shamanic worldview, power and health go hand in hand. If the body is *power-full,* there is no room for illness or disease, and no opportunity for these invasive forces to get in.

When we lose soul, however, a number of symptoms may result. These range from feeling "spaced out" (not really present, or a sense that you are observing life as an outsider rather than engaging with it fully) to pervasive life themes, such as fear, inability to trust other people, depression, and chronic illness.

In the sin-eating tradition also, soul is lost, weakened, or damaged through acts of betrayal—either those we have experienced ourselves (for example, when someone who purports to love us treats us cruelly) or those we have inflicted on others (for example, when we treat someone who loves us cruelly). In such circumstances, our shame or guilt becomes acidic, eroding our souls and causing us to lose spiritual integrity until our pent-up feelings are released through confession or action.

One of the sin-eating arts was "active listening," a way of homing in on the words of patients in the knowledge that the soul speaks in symbols and poetry and will reveal the cause of its own loss or weakness, even if the patient does not consciously know. Expressions such as "I felt like my guts were ripped out" or "I gave her my life" were more than just idle remarks; they were the soul communicating its pain and telling the sin eater that its energy had been lost (and even, in the case of the former, where the energy had been lost from).

Other symbolic expressions suggesting that some deep, intuitive part of us is aware that soul loss has taken place might include the following:

- He broke (or stole) my heart
- I've lost my better half
- I'm nothing without her
- I don't feel whole (or complete) anymore
- I have less energy than I used to
- He was my life
- I'm a slave to her
- I don't have the strength to say no
- I feel as if I'm being torn apart
- I feel like an emotional cripple
- I feel numb
- I feel devastated (torn to bits)
- I feel lost (or trapped or stuck)
- I don't feel anything at all

A lot of these expressions have to do with love and emotional attachments. And, in fact, love can be a form of soul loss too, if we give ourselves so completely and love so deeply that we are no longer ourselves and see our lovers as our whole lives. Our hearts can also "break" when someone we love leaves us—literally break, in some cases. Scientists in Vienna have found, for example, that parts of the heart can enter a "sleeping" state when one suffers emotional pain. Doctors in Russia, Japan, and Brazil have apparently uncovered cases where the heart stops altogether in such circumstances, yet the patient goes on living.[11]

RETURNING THE SOUL THROUGH NATURE

Another way the soul can be lost is by dishonoring nature or ignoring our need for connection with it. Human beings, as parts of nature, *need* to feel their community with the great world soul. Our health, and invariably our feelings of well-being are rooted in this.

This is not a great mystery; it *feels* good to be at the edge of the sea, breathing in the fresh air, to experience the peace of a forest walk, or to sense our place in the scheme of things when we stand in awe looking out on the breathtaking view from a mountain. It is in our nature to be in nature and restoring our connection to it is vital if we wish to remain healthy and well.

In many countries, Japan among them, even today there are roadside shrines (and some in places like shopping malls) where people can rest, pay their respects to the natural world, and receive healing and replenishment from it. The Japanese festival of Obon, which takes place every August, is another way of honoring the ancestors and the spirits of nature. Also known as the Feast of Lanterns, it is a time when the spirits of the dead return to visit living relatives, just as the year is turning and nature begins to rest. *Bon odori* (folk dances) are held, offerings are left at nature shrines and outside people's homes, and colorful lanterns are placed at gravesides. The event closes with *toro nagashi*, when candles are lit inside lanterns and they are set afloat on rivers so the spirits may go free. Another Japanese festival—this one in celebration of life—is Hanami, "flower viewing," where families come together to sit beneath the trees and watch the cherry blossoms open.

In the modern West, we have few such ceremonies or sacred places left. Festivals such as May Day, originally a fertility ritual to welcome the coming of spring, and Midsummer Day, to celebrate the solstice, have lost much of their purpose and meaning. Our nature-inspired rituals have dwindled or disappeared, and our connection to nature is weakened. In turn, our souls, individually and collectively, have become weak, and despite our great wealth and "power," many traditional societies regard us as the saddest people on Earth.

Returning the soul, as you might sense from our discussions so far, often involves the shaman reconnecting her patient to nature, so he is restored to balance and his spirit has a safe, strong, whole place to return to.

In Japan, one method is to accompany a patient (or advise her to go on her own) on a walk into nature to find and make contact with a particular tree that calls to her. She then sits down with her back to it and speaks to the tree of her problems and sorrows. If she listens closely, the spirit of the tree—the great gateway to nature—will counsel her on what to do, while at the same time taking and transforming her pains and giving her power and new spirit in return.

This is not unlike the observation of Doris Rivera Lenz (in chapter 1) that Peruvians will go into nature to shout out their problems to the gods. And in Tuva, too, we find that the patient is advised to take a similar walk and make an offering to a nature shrine, in return for which the spirits will bring back his soul. In all cases, the patient is deeply immersed in nature, at one with the trees and held in the peace of the forest, which is itself invigorating and restful.

In the Andes, soul retrieval is a similar but slightly different practice. Here, the shaman will accompany the patient to the physical location where soul was lost, in order to find and bring back its energy. There is always a physical location where trauma occurred, whether an accident black spot where a car crash took place or a home at the center of childhood abuse, and that is where the soul remains locked. The shaman is able to bring back the soul by negotiating for its release with the spirit of this place and by enticing the soul to return by singing to it

of the joys that await back in the patient's body, now that the trauma
has ended.

In negotiating with the spirit of place, the shaman may also make an
offerenda in exchange for the soul, or simply leave flowers.* If the spirits
of nature are satisfied with the offering and reassured that the soul they
are protecting will be treated well on its return—and if the soul itself
feels loved and safe—it will be released to the patient straight away.

Lenz comments on this Andean practice as follows:

> When a child falls suddenly, for example, its soul can leave its
> body and it may get ill. If this happens, an offering is made in the
> place of the fall, to heal the child.
>
> There are many ways to "call the soul." You can get hold of
> a piece of the child's clothing and make a little doll and decorate
> it with flowers or whatever the child likes, and you call his soul
> in the place where the fright took place. You can also call up and
> use the energies of herbs, a dove's nest, feathers, tobacco, coca,
> or whatever else is needed to help with this healing, but before
> any session, you must first ask permission from Pachamama, the
> spirit of the Earth.
>
> If there is no fixed place where the problem began, then you
> go to the highest mountain or closest river and perform the ritual
> there.

There is another approach to soul retrieval that also works with
flowers, common in countries as diverse as Mexico, Haiti, and Peru.
In these traditions it is believed that the soul can sometimes be, not
lost exactly, but so loosely attached that it is vibrating inside and outside the body at one and the same time. This can happen as a result of
shock, when events that shake our worldviews and undermine all that
we thought to be true can also set our spirits shaking. It is as if we have

---

*We have a vague memory in the West of the connection between spirit and flowers in
our practice of laying bouquets and wreaths on graves or at the scene of accidents where
someone has died. On a symbolic level, we are also negotiating for the release of the soul
and making our offering in lieu to the spirit of the place.

Juan Navarro, Andean San Pedro maestro, standing with his mesa (altar).

Ayahuasca ceremony, led by Amazonian shaman, Javier Arevalo.

Leaves of the chacapa bush. These are dried and bound together to create a natural rattle and healing tool, also known as a chacapa.

Amazonian shaman, Artidoro, preparing chullachaqui caspi by taking shavings from the bark. In the foreground are bundles of guayusa, pinon colorado, and ajo sacha.

Curandera, Doris Rivera Lenz, divining with coca as she officiates in a ceremonial Andean offerenda.

Artidoro, showing the floripondio flower, which grows in an ants' nest and is used magically to generate work and stop people from being lazy.

Leaf walk in Haiti to collect
magical plants for healing
baths. The procession is led by
Houngan Babou, who cracks a
whip to hold malevolent spirits
at bay.

Ross Heaven (right), in ritual
Vodou costume, opens a
ceremony in Haiti. The "mist"
behind the two figures is
reckoned to be a Lwa (spirit)
entering the circle.

Left to right: Buseta (Eva), a female pusanga plant used to awaken passion in the opposite sex. Buseta (Adan), a male pusanga plant used by men for the same reasons. The man rubs the perfume "Adan" on his penis so women become aroused by him. The "Eva" is ground into powder and used by women for the same purpose.

Artidoro, collecting ingredients to make a ceremonial floral bath. He holds "rosa sisa" (marigolds) and a bottle of plant perfume.

Amazonian shaman, Wilson Montez, holds (left to right): mashushillo, floripondio, and buseta, plants that are all used in plant magic.

Wanga (charm) bottles hanging in a Haitian Makaya House (house of magic). The bottles contain plants and essences to attract good fortune to the person they are made for and remove evil magic or negative energies. Also note the snakes: in plant spirit magic, many shamans recognize a primal connection between serpents and plants and the affinity they have for each other.

The root of the jergon sacha tree, which is used in the snake medicine.

Javier stands next to a jergon sacha tree. The plant has the same markings as the bushmaster snake and is used as an antidote to snake bites.

Day breaks over the Amazon.

Howard Charing holding lengua del perro leaves. Lengua del perro ("dog's tongue") is so called because it resembles a dog's tongue. By using it (in pusangas, etc.) it is said that a person will be made loyal and faithful.

nothing left to hold on to, and all of our balance is gone. Shocks like these can lead to trauma, but if the soul is caught quickly enough it can be healed before deeper wounding occurs, by forcing it back into the body and stabilizing it there so that balance is restored.

One method of doing so is to swaddle the patient tightly in sheets or blankets so that the soul is pressed back into the body and held there.* Inside the blanket are placed flower petals, and these may also be sprinkled on top of and around the patient's recumbent form. As the patient lies in her sweet-smelling cocoon of flowers that soothe the soul, the shaman will sing to her in lullabies and whispers of how beautiful the world is and how much she is loved and wanted by her people. Flower perfumes and essences may also be sprayed over her, their aromas anchoring her memory of the sweet words she is hearing and the prayers for her soul to the spirits of nature. Then she is left there for a while in the gentle heat of a rising sun, before the shaman unwraps her and welcomes her home into a new possibility of life: a rebirth through flowers.

Another method is that related by the Mexican medicine man, don Abraham, who speaks of the *alta miza* herb, which is used to gently beat the patient to "heal traumas, to make a regression . . . Alta miza grasps your spirit and moves you backward . . . until you reach the place which hurts. And then she confronts you with the pain. And she will heal the pain."[12]

This is similar, in turn, to the Amazonian use of the chacapa to remove negative energies and restore spirit to a patient (as we saw earlier in this chapter). In both cases, it is the plants, directly, that offer the healing. It is interesting, as well, that alta miza is feverfew, which has long been respected for its properties of healing and purification, and which was widely planted in old England in the belief that it would purify the air and prevent the spread of plague. Gerard said of it that it "cleanseth, purgeth or scoureth, openeth, and fully performeth all that bitter things

---

*This may also be the origin of the practice of swaddling babies, traditional people recognizing that the soul of a baby is less attached to its physical body and needs to be held in place until the child has "grown into itself" and become established in its body.

can do."[13] (Also recall Loulou Prince's comment on bitter herbs as a means of purifying the blood and rebalancing energies.)

We know from the work of Backster and others (chapter 1) that plants have an affinity to human beings, that they know our pain, and that their intention is to love and to heal. Simply being close to them and their energy fields can be enough to call back the soul.

CASE STUDY:
SOUL RETRIEVAL WITH PLANT SPIRITS

When lost soul parts are returned to a patient, old memories, sometimes long forgotten, may also resurface because the energy that held them, repressed due to the shock or trauma of the event, is now back. These memories can be part of the healing process and may lead to new revelations in their own right.

Howard relates an experience in which a message from the plants opened a gateway to a client's hidden memories. This, in turn, led to the police reopening a forty-one-year-old murder case.

> Following a soul retrieval session, Joanne called me in consternation, as certain memories were resurfacing and causing her discomfort. I took a shamanic journey for her [see chapter 1 for a discussion and exercises using this technique] and found myself in the town of Bradford in the north of England, on one of those old-fashioned streets. Around me was a strong scent of lavender, which I recognized as wood wax, a smell I hadn't experienced in years. I breathed it in and my attention was drawn to a small nearby shop, which I entered. The source of this specific lavender scent was the polish on the wooden floor and cabinets. I was also struck by an aroma of herbs and spices.
>
> In the shop was a thin man, but as I looked closer I could see that it was a woman disguised as a man. I followed her into the street, and outside I saw the child Joanne, whose hand was being held by her grandmother. Her grandmother had the manner of someone who was acting as a "look-out."

All this time, I was carrying out a simultaneous narration of the journey, and when I returned to the here and now, Joanne was in a state of sadness and relief. She told me she knew my journey was true when I described the particular smell of that old-fashioned lavender-scented wax and the spices. She was aware that this event had happened—that her mother was the woman dressed as a man, and that she had gone to this shop to kill the owner.

A few days later, Joanne called to say that more memories of that time had resurfaced. She remembered the motive for the murder: the shopkeeper was a moneylender, and her mother was in serious debt to her. She also told me that she had gone to the police, who had reopened the case, based on what she had told them. Since her mother was terminally ill, however, and the case could not be effectively pursued, the police had decided not to press charges.

Joanne had gone to see her mother, who was close to death. Joanne said to her, "Mum, I know you did it, I know you murdered her." Her mother could not speak, but she began to cry. Her secret was out. This was a true deathbed confession, which at the very last moment reconciled Joanne with her mother. Shortly after this, her mother went into a coma and died.

The whole story moved me. The beauty of it was that Joanne could at last put this gnawing, hidden secret to rest and move on in her life. I was also touched that her mother could finally unburden herself, at the moment before she died.

One of the magical qualities of this soul retrieval was how the spirit of lavender reached out and drew me to where I needed to be so this healing could take place.

As a concluding note, Joanne later sent some newspaper clippings about the murder, and I was shocked when I looked at the photographs of the shop—it was exactly as I had seen it in my journey.

Soul retrieval is not a one-off act; I see it as the start of a process that leads to the integration or union of life force. From that place, a person can move forward in her life with renewed vigor and passion.

PROTECTING OUR SOULS AGAINST LOSS

To receive initiation truly means to expand sideways
into the glory of oaks, mountains, glaciers, horses, lions,
grasses, waterfalls, deer. We need wilderness
and extravagance. Whatever shuts a man away from
[these things] will kill him.

ROBERT BLY

One action that every spiritual system advises in order to strengthen and protect the soul is time spent alone in nature. In some cultures this has been refined into a psychospiritual process known as the vision quest or the sitting out.

In essence it is simple. The student walks out into nature to find a place where he or she will not be disturbed for a period of time—whether for hours (often overnight, in the form of a night vigil) or days (four days is common)—and commits to staying there, no matter what.

The student enters this place with a question in mind: *Who am I? Where should my life take me now?* and so on. He takes with him a minimal amount of equipment from the material world—perhaps just a bottle of water—and will usually have received a cleansing (a sweat lodge ceremony or purification by smudging) and undertaken a fast before his pilgrimage begins.

The journey into nature is a walk of attention, so he allows the perfect vision-place to call him. When he finds it, he will smudge the area and form around himself a protective circle of stones. Then he will sit quietly, observing nature for the answer to his questions, in the knowledge that nature is alive and talking to us, and that if we only slow down to its pace, we will find all the answers we need.

There is no note taking and no other distracting of the self. The student simply sits and watches for omens. The shape of a cloud, the flight of a bird, the midnight visit of a fox, the wind in the trees, the smell of the pines—all of these can provide answers if we free our minds to receive them and let them speak to our souls.

For those of us raised in the hectic pace of Western culture, with all its distractions from the soul, our first response to this self-imposed regime of sitting in stillness is often irritation. We feel we need to be *doing* more than we are—instead of *being* who we are. Such irritation is healthy—it is the spirit freeing itself of toxins and the mind relaxing into a less artificial pace.

Then, after a time, the student enters a state of oneness with nature, his eyes open, and he sees the truth of who he is and what the world is truly about.

This is how Richard, one of Ross's students, described his quest for vision:

I have recently completed my first Vision Quest and thought I would share some of my experiences and thoughts.

Firstly, I just seemed to "know" that it was going to happen on a certain date, although the weather either side of it was wet and cold (good old British weather!) and I was left wondering if it was the "right time." I just had this inner feeling that it was right and to trust that it was going to happen, almost like I had made an appointment, and it was also involving so much more than just me, I had no choice but to now keep it.

I also found the fasting easier than expected. Now, I like my food, and have never gone without for very long, but it seemed easy and helped me to focus and plan for my quest.

Now, the place that I selected is on the top of a hill in the middle of the Mendips, quite remote for England, exposed to the weather but a fantastic location to sit and think, and it was my number one choice for the night.

I built my circle of stones and smudged myself and surroundings and settled down. After quite some time in wriggling around and settling myself, I started to slow myself down to the heartbeat of the planet. The one feeling that still remains with me is of my connection to Earth itself.

I find it quite hard to explain, but, from being still up there and watching the sun set in the West and knowing that I was going to see it rise in the East, I swear I could feel the Earth turning in space, with me sitting up there on my hill. The feelings of connection became very strong, and I could then see just how affected by the elements we are, the pull of the moon, the sunlight on our bodies, the rain and clouds, everything.

It was an amazing thing for someone who lives in a society that operates twenty-four hours a day with no changing seasonally or with nature anymore, removing ourselves further and further from the natural ways. It was very enlightening for me and I have taken that lesson with me, and I am now trying to be aware of the moon's cycle and the seasons/weather and how it affects us all and how we feel.

I also got some messages from the wildlife around me that confirmed my questions that I came out with and when I came down from my hill at the end of the Quest, I felt connected once again with the pulse of the planet, the real clock who I have been removed from for so long. It was an amazing experience and very powerful for me and still is two weeks afterwards as I reflect.

There is no greater protection for the soul than sitting out in nature. Through the power of our experiences and realizations there, we see how trivial our petty human concerns are and how meaningless the jostling for status and control within our cities has become. And once we see this, we understand how we can make other choices, so we never put our souls at risk. Nature, the real world beyond the shadows, will teach us if we listen.

> *Teach us to care and not to care*
> *Teach us to sit still*
>
> T. S. ELIOT

The following exercises are an introduction to healing the soul.

Removing Intrusions in the Body

Many shamanic traditions believe that all of our body parts have a specific spirit energy to them. And all of these individual spirits, together, are what make up the soul. The Sora Indians, for example, say that our life force is born in the heart and carried by the blood so it touches all parts of the body.* For the Wana of Celebes, our spirit energies are in the fontanel, the liver, and other internal organs. In Mongolia, each body part has its own *ezhin* or spiritual ruler, while, for the Cuna shamans of Panama, all body parts have a spirit. Thus, in all of these traditions, a physical ache or pain is a message from the soul, bringing us information about more deep-seated spiritual or emotional problems that we are carrying within us.

If you relax with your eyes closed and scan your body, it may be that you become aware of a limb, an organ, or another body part that calls for your attention. If so, journey into that part and hear the message behind this pain. What is it trying to tell you? What is the emotional content beneath the physical ache? This is the message from spirit about the way life has affected you. The spirit may have suggestions, as well, as to how you can change and avoid such pains.

This pain-energy is one form of spirit intrusion. In your journey, see it as an image-being, and note how it appears to you and what it has to say. This is useful information that will take you to a deeper level of understanding.

There will also be a time and a place in which that intrusive energy first entered your body. When and where was it? If you can, visit this place in physical reality (if not, journey to it using your active imagination) and, when there, ask that body part how you should remove this energy now that you have heard its healing message.

The spirit-being may have a request, perhaps a ritual or ceremony such as an offerenda. Or you may simply be told to use your breath to

---

*An interesting connection of this belief to recent research is in the fact that about 65 percent of the cells in the heart are *neural* cells. According to Stephen Harrod Buhner in *The Secret Teaching of Plants*, "They are the same kind as those in the brain and they function in exactly the same way . . . they also have direct connections to a number of areas of the brain, and produce an unmediated exchange of information with the brain." In other words, we sense and understand the world with our hearts and emotions as well as (and, in fact, before) rationalizing it with our minds.

blow this energy into the Earth. Whatever the case for you, when you have performed this healing action, thank the intrusive energy for the information it has given you, and leave flowers at the spot you return it to.

When you reach your home again, take a warm bath with three spoonfuls of salt and one fresh lime chopped and added to it. In almost every culture of the world, salt and lime are used for cutting through magic and will remove any residual energies and pain from your body. If you have them, scatter marigold petals on the surface of the bath water as well, to invigorate your soul.

Returning Soul Energy

This is an exercise to take back energies you have lost and to find reconnection to the Earth.

Close your eyes and, in a quiet frame of mind, see yourself standing on the threshold of a great forest. When you step into it, you will find yourself in a passageway of trees, at the end of which is a source of light. This is the natural daylight of the otherworld. Walk toward it and enter the realm of nature.

Look around in your environment for a river, a stream, or other body of fresh, running water. Follow this back to its source, and there you will find your plant ally in the form of a spirit teacher. Ask that he or she guide you to the place where your soul parts, lost over the years through the impact of life on the spirit, are waiting to return to you.

When you meet these soul parts, they may appear as symbols, or as sensations, or in the form of children or younger aspects of yourself, and you may see the circumstances leading to their loss. Ask them about themselves, what has led them here, and what actions you can take to ensure that they are safe and happy and will not leave you again. If you are inspired to, promise that you will take these actions of self-healing and protection. Then, when you are ready, hug these soul parts to you and return, via the route you have just taken, back to ordinary awareness. With your arms folded across your chest, feel their energy flow into your body.

For the next few days, drink teas (morning and night) made of gentle herbs such as chamomile and hops, and watch for any unusual experiences, including reveries or dreams that have meaning for you. The energy you have just returned to yourself has its own memories and consciousness and will communicate these to you, giving you insights into yourself as you become more whole again. You may also wish to use a bila or dream pillow (see chapter 2) to amplify these messages from your soul.

Releasing the Soul Parts of Others

We may also be holding on to the soul parts or energy of others even though they are no longer a part of our lives. This can happen, for instance, when a love affair ends: even though our lover has gone, it can still feel as if he or she is present and "haunting" our thoughts. We continue to behave in a way that would please (or hurt) our lover, even though we have nothing to prove anymore. It can also happen when someone we have been close to dies and it is if a part of ourselves has died with them so we are in permanent grief, and, again, our lives are not fully our own. In this way, we give others a part of our power and their energy still guides and controls us.

To explore this, relax and journey back to the forest (as in the preceding exercise) to meet with your spirit ally and ask that he or she show you the energy of others that is still a part of you. This may appear as an image of the person, and you may be surprised at the soul parts that are revealed as you meet this person again who, consciously, you thought you had long let go of.

Now that you see this energy, speak with your ally and ask to be shown the circumstances in which you feel drawn to take power from another person or to let them into your soul. What is the need in you that causes this? And how best can you control this tendency so you draw more beneficial energies from other sources?

When you are ready to return, gather these soul parts in the same way as before and come back to ordinary reality.

To return this energy that you have been holding on to, write down the name of the person this energy belongs to. Put as much concentration and intention as you can into the writing, so that energy as well as ink flows from your pen to the paper. When you are done, fold the paper with their name inward and write across it, "I release you," then wrap it in flowers, tie it with ribbon, and take it into nature, where you can leave it at the base of a tree or cast it into flowing water. Once this is done, turn around and walk away without looking back.

The Quest for Vision

Simply put, the vision quest is a method of strengthening the spirit through connection to the Earth and the natural order and flow of the universe. Begin it with a question in mind, something you will allow the spirit of nature to clarify for you. Any question you need an answer to is fine. Then commit to a time and a place in nature where you will undertake a vigil to find the answers you seek.

Arrive before sunset on day one and purify your space. Remain in this spot until sunset the next day. Don't take anything with you that might distract you. Instead, pay careful attention to nature and your own internal processes as you sit quietly with an empty, open, and receptive mind. When it is time to go, clear your space, leaving it exactly as you found it, and thank the spirits for being with you, before you return to your home.

Reflect on the information you have received and watch your dreams for the next week or so, recording them in a dream journal (again, you may also wish to use a dream pillow to assist your dreaming process— see chapter 2). At the end of this time, make your notes.

What have you learned from your quest and how does your soul feel now?

5

PUSANGAS AND PERFUMES: AROMAS FOR LOVE AND WHOLENESS

Of all the senses, none surely is so mysterious as that of smell . . . its effects upon the psyche are both wide and deep, at once obvious and subtle.

DAN MCKENZIE

Beautiful aromas derived from flowers and herbs have always been used for healing, as far back through history as we can trace. Even the word *perfume* comes from the Latin *per fumer,* "through smoke," a reference to its ritual use in ceremonies to meet the gods of health and prophecy.

Smell is the most powerful of human senses. Many people can remember smells from their childhood, fifty or sixty or more years ago, and scientists have shown that even a year after people encounter a new aroma, 65 percent of us can still recall it accurately. By contrast, visual memory drops to 50 percent after just three months.

Because olfaction is handled by the limbic system, which also controls our emotions, smells evoke feelings as well as memories, so we experience not just an odor but a mood associated with it. Even if we do not recognize a perfume and could not name it, we can respond emotionally to it.

And yet fragrances, despite their qualities and evocations, are invisible, ephemeral, part of the unseen world. For these reasons, shamans regard them as spirit beings, able to create feelings, change moods, alter an outlook on life, deliver healing, and change a person's luck, as well as bringing tangible and practical benefits. There are scents to bring back lovers, for example, or make you rich, to win court cases, or turn you into a more successful and potent lover. All of these formulas and effects are within the skill set of *perfumeros* (shamans who work with fragrance), who have made allies among the healing and aromatic plants.

FRAGRANT FASCINATIONS

Our fascination with perfume began many thousands of years ago, with the burning of scented plants mixed with gums and resins to create incense that was used for practical as well as for ritual purposes—for merging with the natural world to increase the effectiveness of hunting, for example, while also for calling "the owner of the animals" to ensure plentiful game and protection on the hunt.

Anthropological evidence shows that from around 7000–4000 B.C.E., our ancestors were combining olive and sesame oils with plants and flowers to make the first ointments. Some anthropologists speculate that early hunters, having covered their bodies with the scent of fragrant plants to mask their smell and attract game, noticed the healing properties of the plants they used and their curative effects on wounds sustained in hunting, and this is what led to the formulation of ointments and balms. Others believe it was the women who first began to explore the effects of different fragrances as they met them in the plants they gathered and worked with. Which of these came "first" is immaterial, since both these paths of learning informed our progress in the use of fragrance.

We know from historical sources, that by the year 2697 B.C.E., a body of knowledge on plant medicine was certainly well established in the East, and we read in *The Yellow Emperor's Classic of Internal Medicine*,[1] for example, of many uses for scented herbs.*

As for the West, David Hoffmann, in *Welsh Herbal Medicine*, discusses the preparation and folk use of *meddyginiaeth*, "natural medicines" in Wales, by the Gwyddoniaid—men of knowledge who were ancestors of the druids. This evidence stems from at least 1000 B.C.E. By 430 B.C.E. in Wales, the Druidic text, the laws of Dynwal Moelmud show that plant medicine had come to be highly regarded and was protected and encouraged by the state, with commerce, healing, and navigation known as the "three civil arts." [2]

One of the interesting folk uses for fragrant herbs within these Welsh traditions was the practice of "burying illnesses" beneath aromatic plants. The sin eater, Adam, for example, would lay out wooden stakes in his garden, beneath which he would bury an animal bone with the name of a patient scratched on it. He would then plant certain flowers or herbs on top of these "graves," according to the nature of his patient's illness: thyme for colds and fevers, for example; rosemary for lethargy; parsley to purify the blood; and marigolds, among their other more spiritual virtues, to ease skin complaints and inflammations.

All of these plants might today be used by an herbalist to cure the same ailments, either in a tea or a salve, but in this folk practice, it was the energetic or sympathetic connection between plant and patient (represented by the name on the bone) that mattered. Each morning, Adam would walk his garden, whispering to the plants and crushing a few of their leaves between his fingers. As they released their aroma, every day this emanation carried away a little more of the illness, until the patient was cured. As in all shamanic practice, these plants were regarded not as medicinal substances but rather as spirit allies who brought healing to the body.

Chinese Taoists believed that a plant's fragrance *was* its soul, a belief later endorsed by the Gnostic Christians of 100–400 C.E., for whom fragrance was the spirit of the plant and a gateway to the greater soul

---

*This ancient classic, translated by Ilza Veith, is available in a new (2002) edition from the University of California Press.

of the world. In their ceremonies surrounding death, the corpse was washed in perfume and incense was burned around it so the soul of the deceased would mingle with these fragrances and, through them, find its way to God.

It is, however, the Egyptians that are most associated with perfume and who left the most evidence of their fascination with the mystical attributes of scents. Manuscripts such as the *Papyrus Ebers* (1550 B.C.E.) describe the use of plants such as elder, aloe, cannabis, and wormwood. Others, from even earlier, record the use of herbs in temple incense, oils, and salves. The Egyptians used cinnamon to anoint the bodies of the living, for example, and myrrh—considered more precious than gold—to embalm the dead.

Wall paintings such as those in the temple at Edfu show the distillation of perfume from white lilies. Others depict the use of aromatic cones called *bitcones* as adornments for the heads of temple dancers. These cones would melt into the hair and release their fragrance as the maidens danced for the pharaohs and gods. The Egyptians consecrated their temples with incense cubes made from scented plants, gums, and honey. The earliest known use of perfume bottles is also Egyptian and dates from around 1000 B.C.E.

Another use for aromatics was in fragrant sweetmeats called *kyphi,* which means "welcome to the gods." This mystical substance was eaten in the temples of Ra to induce states of trance. Through the audience with the gods this brought, healing dreams would result, which were said to be the most potent cure for grief and a comfort to the soul.

We still have the recipe for kyphi, thanks to Dioscorides (40–90 C.E.), a Greek physician who studied medicine in Egypt and produced a five-volume work, *De Materia Medica,* on the "preparation, properties, and testing of drugs." The ingredients, mixed more-or-less to taste, are chopped raisins, honey, sweet flag, rose, lemongrass, cyperus tuber, myrrh, amber, and wine. These were added one at a time (starting with the raisins) and allowed to steep before the mixture was boiled to a paste, rolled into balls, and ritually eaten.

The knowledge and understanding of plants and herbs, and not just of fragrances, was extremely well developed in Egypt. This is little recognized by modern medical science, even though some of its "own" techniques rely on it. One example is the pregnancy test. Egyptian herbalists were able to determine pregnancy as well as fetal gender by soaking small bags of wheat and barley in a sample of the woman's urine. The growth of both plants was accelerated if the urine contained pregnancy hormones. Further, if the barley sprouted, it meant a female baby; if the wheat sprouted it signified a male. By contrast, modern scientists did not "discover" the urine pregnancy test until 1926, and they did not "develop" the wheat/barley test until 1933.

THE WORLD OF MAGICAL FRAGRANCE

But the use of fragrance to engage the gods was not restricted to China and Egypt. Through experience, quite independently of one another, a number of cultures evolved the conviction that beautiful smells provide a doorway to another world.

The Hebrews used fragrance in their religious ceremonies and to initiate priests, their anointing oil consisting of cinnamon, myrrh, and calamus, mixed with olive oil. The ancient Greeks believed that perfume was god-given and that sweet aromas were how the deities made their presence known. They used the word *arómata* to describe the use of fragrance, making no distinction between medicinal and mystical perfumes, between incense and medicine, or between spiritual and pragmatic uses. Every plant contained magic. Bay, for example, was a staple of Greek cooking, but was also used by the oracle priestesses of Delphi, who would sit within its smoke, heads covered, to enter the otherworld and allow the spirits to speak through them during their divinations. In India, too, in ceremonies of prophecy, seers called *dainyals* would cover their heads with cloth and surround themselves with cedarwood smoke, the aroma of which would send them into trance and chanting.

And fragrant plants were being used extensively throughout Europe. In the Middle Ages, Hildegard of Bingen (1098–1179) was an ambassador for the connection between religion and the healing spirit of the plants. As well as an Abbess, Hildegard was an herbalist and is credited with the invention of sweet-smelling lavender water, which she saw as truly divine.

"Carmelite water," also developed at this time, offered a "miracle cure" for spiritual diseases such as melancholy (regarded as a form of soul loss) and for improving the powers of mind and vision. The monks who produced Carmelite water guarded its spiritual formula, but we now know it was based on melissa (a plant regarded as a "spiritual communicator") and angelica (which was equally effective against evil spirits and infectious diseases, both of them forms of spirit intrusion—see chapter 4).

Another plant with a spiritually protective purpose during the Middle Ages was rue, which also bestowed "second sight." Indeed, rue was believed to be so powerful against conditions such as soul loss and melancholia that it was named from the Greek word, *reuo* ("to set free") and was used in many spells and formulas devised by the Welsh sin eaters, who knew it as *gwenwynllys* and used it as an antidote in cases of spiritual as well as physical poisoning.

It was France, however, which emerged as Europe's leader in the therapeutic use of fragrance. The term *aromatherapy*, in fact, was invented in 1928 by Rene Maurice Gattefoss, a French chemist whose interest in essential oils was stimulated when he burned his hand in a laboratory accident and plunged it into a pot of lavender oil to cool the burn. It healed within days, faster than it would have with any other treatment available at the time. Gattefoss was inspired and began to experiment with essential oils and fragrances from that day.

Gattefoss also inspired others, including Jean Valnet, a French doctor who worked as an army surgeon in World War II and found essential oils such as thyme, clove, and lemon to be just as effective in treating wounds and burns. He later extended his work with fragrances, using them with equal effectiveness to treat psychiatric problems. Valnet went on to write a book of his discoveries, which was called *Aromatherapie* in France and *The Practice of Aromatherapy: A Classic Compendium of Plant Medicines and Their Healing Properties* when it was translated into English.[3]

Today there are over 20,000 commercial fragrances on the market, and the number of new releases each year has increased by more than 400 percent since 1973. The age-old associations between pleasant smells, a healthy soul, and the visionary call of perfumes to and from the gods has not been forgotten, however, even in these times. International Flavors & Fragrances, Inc., is a leading manufacturer of perfumes, with an income of around $300 million a year. And yet Sophia Grojsman, the perfumer for this company, still says that her inspirations for new perfumes come from dreams, images, ancient cultures and practices, and the world of nature. When creating a perfume, she says, "you must always have an image in your head." Her intention, as with all shamanic healing, is to produce balance: "Perfumery is closely related to music. . . . A fragrance that is not well balanced is not well accepted."

PUSANGA, THE FRAGRANCE OF LOVE: AN INTERVIEW WITH TWO PERFUMEROS

> . . . *a certain solid fragrance, risen from the earth, lives*
> *darkly in my body. I love you without knowing how, or*
> *when, or from where . . . I love you because I know no*
> *other way.*
>
> PABLO NERUDA

Fragrance, of course, has also been long associated with the art of love. In Japan, geisha girls priced their services according to the number of incense sticks consumed during lovemaking, while in Indian tantric rituals, men were anointed with sandalwood, and women with jasmine, patchouli, amber, and musk. Saffron was also crushed and smeared beneath the feet. In Europe in the 1700s and 1800s, the use of eau de cologne became a widespread and fashionable trend; the morning ritual in many homes began with its application before a suitor of either sex would call upon a lover. This blend of rosemary, neroli, bergamot, and lemon was also used internally, mixed with wine, eaten on sugar lumps, even taken as an enema, to refresh the "inner self" and cleanse the spirit so that lovers could meet each other with a "pure heart."

But it is, perhaps, in Peru, that the magic of perfumed love has reached its highest skill, in the formulation of *pusanga*, which is often referred to as the "love medicine of the Amazon," although it is far more than that. Specialists in the use of fragrance to change luck and attract good fortune are known as perfumeros. One such specialist is Artidoro Aro Cardenas. Another is Javier Arevalo, an ayahuasca shaman we have met in this book before, and who also works with fragrance.

Artidoro, how did your involvement with perfumes begin?

AAC: The story of my path of medicine began when I watched my brother-in-law, who healed and chanted. I didn't have any profession because my parents were poor. I only had studies up to fifth year primary and I wanted to get out of the place we lived, Parimarques on the River Ucayali, a day downstream from Pucullpa.

I felt trapped in that pueblo with all my brothers, poor, drinking too much, earning next to nothing. So I fled from there to the Montana Tamaya, Alto Ucayali [Upper Amazon jungle]. I was just seventeen years old.

I used to watch how the curanderos there worked. I loved listening to what they talked about, how they prepared their remedies, their canticos.* Then I went off on my own deep into the jungle, to know the plants little by little, to smell the leaves and roots of all the different medicines. I had no maestro to learn from so I dieted plants for a year and a half alone, and then I returned to the city. I used agua florida, timalina, camalonga,† and dedicated myself to studying all about smells.

How do you use perfumes to help people now?

AAC: I get people coming for help with family problems where the woman has gone away from the man or the man has gone away from his children.

Canticos are chants, similar to icaros.

†Amazonian plants used in perfumes.

Supposing the woman has gone off, I use pusanga to bring her back so that the family can consolidate again. I call the plant spirits which work for that—pusanga plants such as renaco, huayanche, lamarosa, sangapilla— and I call her spirit back to her home.

Or let's say the mama is here with me and the father is far away. I pull him back so he returns to his home. In a short time he will be thinking of his children and his wife, and he comes back.

I don't need to have the actual plants in front of me, I call their spirits. I make my own perfumes from plants, no chemicals. They have wonderful smells. And I chant at the same time as I rub them on the children and the woman. Then the man starts thinking or dreaming of them.

How does perfume magic like this work?

AAC: *A smell has the power to attract. I can make smells to attract business, people who buy. You just rub it on your face and it brings in the people to your business. I also make perfumes for love, and others for flourishing.**

These plants are forces of nature; they contain spirit. I watch for what that spirit attracts: maybe bright birds or butterflies, maybe many different animals come to feed from it. A plant that draws bright birds will also draw beautiful women; a plant that is popular and has many "customers" will also be good for business. So these are the plants that I use to help my patients.

Javier Arevalo tells a similar story of humble beginnings. Several generations of his family have been shamans, and at the age of seventeen, Javier knew this would be his future too; but it was not until he was twenty, when his father died from a *virote* (a poisoned dart from the spirit world) sent by an *hechicero* (sorcerer) who was jealous of his father's powers that Javier felt compelled to become a shaman.

His first instinct was to learn the shamanic arts so he could avenge

---

*Flourishing is a term often used in Amazonian and Andean plant medicine. Translated, it means something like 'growing in luck and success' through the attraction of good fortune; as a plant will flourish in good soil and sunlight.

his father. But his grandfather convinced him that this was not the solution, that the only way to defeat evil was to spread more goodness in the world. Javier took the message to heart and found solace in the plants instead.

Javier, how did your involvement with healing begin?

JA: My grandfather saw that I was bitter and told me that it would not get me anywhere. My heart was still hard and I wanted to kill! Bit by bit, though, through taking the very plants that I had intended to use for revenge, I learned from the spirits that it was wrong to kill, and my heart softened.

A shaman learns everything about the rainforest and uses that knowledge to heal his people, since they do not have money for Western doctors. The spirits or plant doctors come to me and say that they will cure a person if he takes a particular plant. Then I go out to look for the plant. It is said that every environment has the plants to heal the people.

As part of his apprenticeship the shaman spends years taking plants and roots, each time remembering which ailment is cured by what. The maestro goes with the apprentice into the wilderness and gives him the different plants and it is like a test or trial to overcome. One plant may cure lots of ailments.

You are respected as an ayahuascero, but you are also a perfumero. How do you use perfume magic?

JA: Through my work with the plants, I have learned how to make pusanga, the Amazonian love potion. Pusanga has the power to attract anyone you wish, for the purposes of love, sex, or marriage.

Take the case of a woman who refuses when you offer her a Coca-Cola because she thinks you are lower class and that she is better than you. That makes you feel like rubbish so you go to a shaman and tell him the name of the girl. He prepares the pusanga. Three days go by without seeing her and she begins to think about you, dreaming about you, and begins looking for you . . .

In the West, people often look down upon such magic as manipulative —they may even see it as evil because it takes away a person's choice and free will, so he or she has no option but to love you. In the Amazon, however, it is considered normal practice to use pusanga in this way. And, in fact, despite Western morals around the issue, when it comes right down to it, in the United States and Europe, people are often willing to use love magic to find or return a lover as well. Once we get past the "ethical" considerations, we can be just as "manipulative" as the people of Peru.

Ross relates, for example, how he is approached for pusanga by people who even provide a detailed "shopping list" of the qualities their man or woman should have when the fragrance draws this lover to them, like this one from Judy, in 2005: "I would like to bring in someone who is very kind, generous, loving, loyal, near my age, likes travel, country walking, and has a nice speaking voice. He should have lovely qualities and be successful and professional at work, a good sense of humor, make me laugh, and treat me like a queen or princess. He should enjoy travel and discovering new adventures. He should be my best friend and someone that I adore too." It seems that no matter what our stated morality on this matter, we have very clear ideas where love is involved.

Perhaps, like Judy, the people of the Amazon are in general more honest and upfront about their needs? Or perhaps they carry a less Christianized concept of "right" and "wrong" so are less afraid to ask for what they want? We asked Javier to comment on the moral question:

JA: Yes, we shamans understand there is an ethical concern, but put it this way: what if it happened to me? Let's say I found a woman ugly and she did pusanga magic to make me marry her. Of course, if I found out I'd be outraged and it would be awful if I only discovered it after having children and making a home with her!

But the truth is, I would never know! I would be hopelessly in love with her, and because I had seen beneath her physical appearance, into her soul or her personality, my love for her would be genuine and deep! She would be the mother of my children! My wife! So the pusanga has not taken away my freedom, it has given me more: it has freed me from my prejudice and let me find real happiness.

That is also why pusanga is a secret. You should never tell someone you have used it on them, otherwise its work is undone.

But [we persisted] does anyone have real freedom if everyone is using pusanga?

JA: Does anyone have freedom anyway? We are all taught what to believe, what is right and wrong, from when we are little. Are our minds really free? Pusanga is just a different freedom.

But we all like to think we are free. If people are using pusanga on us, though, surely we become slaves to their will and victims of magic?

JA: [Laughing] You think you are not subject to magic every time you are with a woman or, if you are a woman, with a man? You think the woman you met tonight at the dance wears the same pretty dress every day, the same makeup, the same scent, when she is scrubbing the kitchen or at her factory job? You think that man dresses in a smart suit or wears that expensive aftershave when he is working in the fields? No!

They are doing those things to present themselves in a certain way, a way which is more attractive, but obviously not always true! We all use magic every day in order to make people like us and get what we want. Pusanga is just another way. Underneath everything we are all just looking for love.

As if to prove his point, a few days later Javier asked the group of Westerners we had taken with us to the jungle what they wanted from their lives. Many of them at first gave "cosmic" and "spiritual" answers to do with putting the world to rights, resolving planetary issues, saving endangered species, speaking with the flowers, and so on, and were quite mute when Javier spoke about pusanga and its ability to meet their personal (rather than planetary) needs.

After time for reflection, Javier asked again what our participants really wanted, and this time they admitted that what they wanted, behind their desire to save the world, was love. A personal love in their own lives.

So why had they not said so in the first place? Many replied that it had not felt okay for them to ask for love. This was the message they had heard from their mothers and fathers ("Who do you think you are to ask for such things?" "You've had more than enough!"), from teachers, and from the church ("Do unto others [but not unto yourself] as you would have others do unto you"); and through this conditioning they now felt their own needs to be secondary. The contradiction, or paradox, was that they believed themselves able to save a planet without first saving themselves—to give cosmic love when they had never received the personal love they needed in their own lives. So how would they even know what this love looked or felt like?

Javier's thoughts on this were simple and enlightening:

JA: If we all had more of the love we need we wouldn't be worried so much about saving the planet! It's because people don't have love that they create the problems of the world in the first place, and why it has to be saved at all!

It would be better if people got what they wanted because then they wouldn't be so destructive. Thoughts tangle up their lives but love solves problems instantly.

THE ETHICS OF ENTITLEMENT

Javier tells a story about one of his own experiences with pusanga, which further illustrates the Amazonian view of the ethics involved in its use:

JA: When I was apprenticed to my grandfather, I was told to practice with the pusanga to gain experience. I began by preparing the medicine without any particular intention, but one day a senorita came to me in my vision and said that to use pusanga effectively I must also learn what love is.

Who was this senorita? She was a mermaid. That's how I can tell you that mermaids exist. She taught me about love and showed me how to prepare pusanga with proper intention and she sang to me, a very sad song. "That is how your lover will come to you, very sad," she said.

Then there was a fiesta one night in the pueblo because it was someone's birthday. I was invited and I took a little bottle of pusanga in my
pocket. I was dancing with all the girls and I saw a girl who was pretty but
she was arrogant and wouldn't let me get near her. She had long finger-
nails.* She said she wouldn't dance with me because I was unattractive.
"Go and find a chola!" [a lower-class Indian], she said, insulting me and
hurting my heart.

"If you don't dance with me tonight, tomorrow you'll sleep with me," I
said, and I went to the toilet and rubbed myself with pusanga. Then I went
back to dance with her and she said "What's the matter with you, damn it?
You know you can't touch me!" She got nasty and wanted to hit me so I left
the fiesta and went home.

The next day I went into the forest to work for three days. On the
fourth day I came out and people told me, "Senorita Suzana is looking for
you."

"Why is she looking for me if she hates me and threw me out of the
fiesta?" I asked. They replied, "No, she has been round every half an hour
asking after you."

When she finally caught up with me she said, "Javier, forgive me for
insulting you in the fiesta." I said "No. You are very arrogant and I don't
want to talk to you." I went home and wouldn't open the door. She waited
outside my house all the while I slept, and she was crying.

I knew it was the pusanga that was working so the next night I said to
her, "You want to be with me?" "Yes, from the depths of my heart," she
replied. "But you are pretty and I am ugly," I said. "No, I want to be with
you," she said again.

That night we made love and she didn't want to go back to her home
anymore. I went again into the forest and she followed me. I went to bathe
in the river and she jumped in too to play with me. She became my girl-
friend for six months.

Later, when the love had passed and the pusanga weakened, she went
back [to her home] as though the whole thing had been a dream. I asked

---

*In the Amazon, this is the sign of an upper-class woman who can afford to grow her
nails because she doesn't need to work.

myself, "How did I actually live with this girl?" but of course I knew the truth: it had been the pusanga working. And the woman had come to me in sorrow, just as the mermaid said.

The pusanga drew her to me but I made her happy and cured her sadness for that time, and she also learned a lesson: that love has nothing to do with status and it is in the power of any of us to love anyone, regardless of their class or even their looks. It is their heart and their soul that matters.

The message of pusanga is that *you can have anything you want* and, indeed, to get what you want is not only healing and empowering for you, but adds to the positivity of the planet as a whole, because if we were all happy and in our power, there would be no conflict or negativity. The world would be healed one person at a time, and even those "manipulated" by pusanga would learn valuable lessons and become happier in themselves. Furthermore, since they would never even know they had been enchanted, no harm would be done. The person would simply be happy and in love.

This message is often lost on a Western audience. In plant spirit workshops in Europe, we can spend an entire weekend debating the issue with participants. Their words may vary, but the resistance to the idea of having what they want is consistent.

One woman dismissed the very idea of pusanga out of fear that if she used it she would attract rapists and men who would abuse her. This, of course, says more about her healing needs and the patterns of attraction in her life than it does about pusanga; for if she could truly have anyone or anything she wanted, why would she choose an abusive man? That is another of pusanga's gifts: just the idea of it forces us to consider the questions, "Why do I believe I *can't* have what I want, or that by getting it, I will make problems for myself?" and, more fundamentally, "What self-limitation prevents me from having my heart's desire?"

The answers often go deep and have to do with early socialization, childhood memories, the shaming we receive from our parents, and even our experiences in the womb. When people are finally able to answer this question for themselves, however, the conclusion is usually simple (though seemingly strange). As one participant put it, "Why can't I have what I want? Because . . . I'm afraid I'd be happy." In the West, it seems, we are so used to having our needs go unmet and living our lives in sadness that we fear we just wouldn't know how to cope with happiness.*

There is another point here as well—one made by Javier—that in the West we are used to control, to being told what to do, what we can and cannot have, in a way that people from the Amazon are not.

JA: *City people [i.e., Westerners and those in Peruvian cities such as Lima that have become Westernized] can't dominate [control] themselves. Even in Iquitos [the jungle town], the poorest people have TV and see the advertisements. It saps their spirit and teaches them that they don't have to look after themselves because Coca-Cola or the government will do it if they just buy the bottle or vote for a politician.*

To control oneself is fundamental to having the strength to work with the spirits, but city people do not take responsibility for themselves and their power wastes away. Now they don't even know what they want or what is good for them or how to get it.

Taking responsibility for, and control of, our lives—bursting through the boundaries of our self-taught or socially learned limitations—is what allows us to find our personal truths and to realize that we can, indeed, have anything we want because that is our birthright. To believe otherwise is to subscribe to a social or individual myth of a little self and the little we deserve.

Once people see this, their perception of themselves and their entitlements often changes radically. One female participant on a plant spirit course in Ireland began the pusanga exercise by saying she had no intention of preparing such a mix. She seemed offended by the very notion of "such manipulative magic." As she walked among the flowers and trees

---

*See *The Spiritual Practices of the Ninja: Four Gates To Freedom* by Ross Heaven (Inner Traditions, 2006) for a discussion of prebirth experiences and how they contribute to our self-limitations, based on a model developed by Ross and Howard. Also see *Darkness Visible* by Ross Heaven (Inner Traditions, 2005) for a study of mystical states that presents a somewhat similar conclusion.

in the garden, though, nature itself spoke to her and told her it was fine to have what she wanted and the plants were only too willing to help her in this.

When she heard this message from nature itself, her opinion changed. She made the pusanga and used it—and she *did* get what she wanted. Her satisfaction from this was much more important, she later said, in creating a better world than all the resistance she had put up to happiness before, which merely spread frustration and negative energy. Her needs had gone unmet, she now recognized, because *she* wouldn't allow herself to meet them. As soon as she took responsibility for herself, however, she—and the world itself—became friendlier and she could act more authentically in it.

"Yes," said Javier when we told him this story. "Plants like people. They feel for a person whose spirit is ill in this way and will say to them, 'I can cure you. Look for me.' Then if that person goes into the forest and looks for the plant, it will cure them."

PUSANGA AND THE SPIRIT OF THE PLANTS

Pusanga is a mixture of plants and roots, each with a particular personality or spirit and each with its own purpose or healing intent. These plants are contained in magical water *(agua de colpas)* collected from clay pools deep in the rainforest where thousands of the most beautifully colored animals gather to drink the healing water, which is rich in mineral content. Many of these animals are natural enemies and would normally prey on each other, but at the clay pools a sense of harmony and peace prevails, as rivals come together to meet a shared need, and there is never any violence as the animals drink. The water itself, then, is a powerful attractant and creates a spirit of balance and cooperation.

Added to this magical water are the barks, roots, and leaves that also have the quality of attraction due to their colors, names, the ways in which they grow, the places they are found, or the sense of power surrounding them. Here again, we see very clearly the doctrine of signatures at work.

Thus, the shaman might make pusangas for good fortune in legal affairs, using plants such as *alacrancito,* a small "pod" root that resembles a tail and is named after the alacran, an insect that has a nasty sting in its tail. Or the shaman may use the roots of *congonita,* a plant with little round leaves and very long stalks that easily become entangled. The magical quality of this plant, reflected in its appearance, is to entangle the opposition in their own words, so their tongues become tied and they are confused and lost in their now-convoluted case, with no possibility of winning.

Then there are the pusangas for commercial success, based on the root of a plant called *mashoshillo,* which looks a little like a string of sausages. "Just look at the form of the plant," said Javier, "like carriages on a train, pulling in the people you need for your business to grow. This plant comes from under the earth, which is clean and pure, so your business will flourish and your affairs will be protected, smooth, and clean."

And, of course, there are the pusanga plants for love, which, depending on the patient's needs, might include *polvo de la buseta,* a powder obtained by grinding up leaves that resemble female genitalia (the male plant looks like the male sex organ). These are used for sexual attraction. The congonita plant, which is used to entangle and confuse a legal situation, is also used for sexual attraction, the principal being the same, but the application different. (Very different chants and intentions will be used to "charge" the congonita, depending on whether it is used for love or legal success.) "This is how couples become involved with each other," said Javier. "With this pusanga, your partner will never leave you."

After immersing the plant mixture in the agua de colpas, the shaman may also add agua florida to the bottle.* The shaman then blesses the entire mix to empower it according to the patient's requirements. She does this by blowing or singing into the pusanga, sometimes with the

---

*Agua florida is a "general purpose" perfume suitable for men and women to use. For pusangas made specifically for a woman, however, and especially for "matters of the heart," the fragrance Tabu is often used instead.

breath, sometimes with sacred tobacco smoke. The traditional blessing or intention blown into the pusanga, which may also be spoken aloud, is *salud, dinero, y amor*—health, money, and love. "This is the way love comes, like a little puff of wind," said Javier, "and that's what goes into the bottle—the breath is the intention for love." Blowing on the pusanga in this way (especially with tobacco) is also said to wake up the spirit of the plants, which will then work for the patient to draw in whatever he most desires.

Once it is made, pusanga can be used like a perfume, with a few drops rubbed on the pulse points of the wrists and neck; or a capful or two can be added to bath water. It can also be used in more magical work. For example, if the patient has a photograph, image, or other representation of something he badly wants, he can set up a small altar and anoint the photograph or other item with the pusanga, then place it on the altar next to a candle he has also anointed. He might also leave an offering to the spirit of the pusanga in return for working with him. When the candle has burned down fully, the spirits will have heard his prayers and accepted his offering, and the pusanga will start to work for him to draw in this thing that he wants.

One other factor is as important with pusanga as with other jungle medicines: the power of faith. You must believe absolutely in the effectiveness of the pusanga and its ability to work for you. In the words of don Eduardo of Cusco, "You must believe without an atom of doubt [because] lack of faith robs your spirit of power"—the very thing which, in one way or another, you are trying to increase or enhance.

Normally a shaman prepares pusanga away from his patients, having first purified his hands with lime or lemon (or occasionally with grapefruit) juice. This serves the dual purpose of cleaning the hands of sweat and salt, which would interfere with the mix, and cutting through magical attachments, since lime and lemon also give a spiritual cleansing. The important thing is to ensure that the pusanga is not contaminated in any way, material or spiritual.

As you will no doubt have realized by now, working with pusangas, as with other plants, is not just a question of brewing up a mixture of roots and perfumes as if from a cookbook. The maestro also needs to be in communion both with the plant spirit and the spirit of his patient in order to charge the energy of the pusanga effectively and achieve the desired results.

Artidoro describes how he works:

AAC: *When I prepare pusanga, I pray and chant the names of all the plants one by one, and I soplada them to increase their strength.* I also speak out the name of the person for whom the pusanga is intended. The client should then use the pusanga as described, and also present themselves to the one they want to attract on three separate occasions. Then they should disappear for two days and after that the person will come to look them out.*

The perfume is important, but the pull comes from the spirit of the plant, which is charged and given strength when I prepare it this way. The client doesn't need to understand the process but she must trust the shaman. Also, she should not allow anyone else to use her preparation; nor should she eat pork or spicy food [while she is using the pusanga].†

The shaman also needs to have skill and he must have dieted the plants to know them because there are many plants and many different pusangas, some with only subtle differences between them. An example of this is amares, which literally ties two people together. Although congonita would entangle the two so they become intertwined, inseparable, this is not the same as tying a person to you, so here I would use renaco instead.‡ Then I call the couple up in my vision and blow on the plants to unite them. The person who asks for this work to be done would also need to be present.

---

*To *soplada* is to blow intention into a plant, person, or object. This may be done with the breath or, more often, with tobacco smoke.

†Also see chapter 2, on the shaman's diet.

‡Renaco is a fig tree that produces masses of complex intertwined branches and roots and various root cuttings might be used in a renaco pusanga to inextricably entangle the lovers. This, again, is an example of the doctrine of signatures: that which naturally intertwines and entangles can transfer this power to others. Another quality of the renaco is that animals from many different species live in its complex root system, all drawn to it because of the protection it offers. No matter what their usual relationships, one to another, all of these animals call the renaco "home" and live in peace there with each other. This is also a "pusanga" connection.

The job, properly done, requires eight days and nightly ceremonies accompanied by healing, relaxing, and cleansing of negative influences. Some people want the tie to last forever, others ask for it to work just for a certain period, or even a very short time in order to make a mockery of a man or woman, get them to come back after they have left, and then make them cry, or reject them. But this is vengeance and I don't agree to doing that.

It is an involved process to tie two people together. For the shaman it requires a high level of commitment because he has to concentrate and incorporate himself spiritually in both of them. Let's say the client is a man who wants his woman tied to him. Then I must call the woman's spirit in the night and blow her onto the man while she is dreaming. When she wakes up she will be thinking about him.

Each night I divine the progress by lighting two mapachos [cigars made from rolled jungle tobacco] together and calling the names of the couple. I puff at the mapachos without inhaling, and as they burn down, I look at how the ashes form. If they merge, the magic is working; if they incline away from each other, there is more work to do. Or the ash might part at first and later join as the couple's spirits meet. When this is happening, the couple won't be able to sleep, but on the eighth night the ashes should merge.

HOODOO ATTRACTION OILS

We find a parallel to the pusanga tradition in Hoodoo, the American folk medicine practice, where aromas are similarly used to create positive change. Hoodoo oils work on the principle that people are highly sensitive to smells and susceptible to the unconscious and emotive effects they produce. They are therefore open to purposeful influence from an aroma if the person using it knows how this scent will affect a mood. With the aroma to work with and the spirit of the perfume on his or her side, the Hoodoo magician can ensure that the needs of his or her client are met.

Over the years, Hoodoo practitioners have developed a detailed understanding of various aromas and have prepared these as formulations to achieve particular outcomes. The examples that follow are fairly standard Hoodoo recipes, but of course, you can modify or strengthen them according to your own needs, by journeying to the spirit of the fragrance or to your plant ally, and asking what else should be added in order to achieve your purpose.

In all cases below, the fragrances (aromatherapy oils are fine) are added to a base oil, such as olive oil, and the resulting scent can be used in the same way as pusanga, by applying it to the pulse points or adding it to a bath. Since they are oils, they can be used in a spray or a burner to freshen a room as well.

The following preparations are used as charms of love attraction and will draw new partners to you or help strengthen existing relationships. In appendix 3 you will find others for legal and financial success and for luck and protection.

"Come to Me" Oil

To attract a new lover, add equal amounts of rose, jasmine, bergamot, and damiana (a drop or two of each) to your base oil, and use it as a perfume whenever you are near the person you desire.

"Come See Me" Oil

Whereas the oil above is a "general purpose" love attractant, this one is used to pull in your "ideal mate." To your base oil add five drops of patchouli and one of cinnamon. Smear this on a white candle, and if you have one, place a photograph of your ideal man or woman next to it. As the candle burns down, visualize and intend that this perfect mate will walk effortlessly into your life.

"Deepening Love" Oil

For this oil, use almond as the base, and to this add seven drops of rose water, seven drops of vanilla (or one vanilla bean), three drops of lemon, and a sprinkling of gold glitter. Use this on the pulse points before going out on a date or add to bath water and wear in the presence of your lover for deepening your romance.

"Sex Energy" Oil

Now that you have your mate, this oil will give you greater confidence and power in lovemaking. It can also be worn to attract new sexual partners. It requires two drops of ginger, two of patchouli, and one drop each of cardamom and sandalwood.

"Love Separation" Oil

To your base oil, add four drops of black pepper, three of vertivert and sandalwood, and one of clove. This is a dual-purpose oil. It can be used to cause the parting of two lovers by introducing the aroma to a room where the couple is together, so that they argue and separate. It can also be used to ease the pain of separation if your own relationship has ended.

THE SECRET OF ATTRACTION

When you pour it onto your skin, the fragrance begins to penetrate your spirit, and the spirit is what gives you the force to pull the people. The spirit is what pulls.

JAVIER AREVALO

The truth about pusanga and Hoodoo attraction oils is, of course, more complicated than a simplistic moral definition of right and wrong allows. Although some might regard these formulas as manipulative, what really gives them their power is not the magic you use on others, but *the magic within yourself*. Self-belief—"without an atom of doubt"—empowers the fragrance to work for you. It is the spirit—*your spirit*—that "gives you the force to pull the people."

What takes place when we use pusanga is not a direct change in other people, but a change in ourselves. We are not so much interfering with the freedom of others or putting a number on them as giving freedom to something within ourselves. It is our *own* power, charisma, confidence, and self-belief that is enhanced and brought out by the sympathetic magic of the plants; *our* charm, not *the* charm that does the work.

The women of Iquitos, for example, are feared throughout Peru for the power of their pusangas, which mean they could steal any man they wanted from another woman. Watching these women of Iquitos walk down the street, they have a radiant, self-assured quality to them. They *know* they are powerful and attractive, and this translates into skills of love, the end result being that they can, indeed, have any man.

Where the real power lies, though, is with—or, rather, *within*—themselves. Because of their belief in the pusanga, their own powers of attraction are increased. The manipulation of others, if it occurs at all, is through their self-confidence, their love of (rather than shame at) their own bodies, and their relaxed sexuality. They are happy and comfortable in their skins in a way many Westerners are not. With the spirit of the plants on their side, the women of Iquitos are able to be all that they naturally are, and their inner beauty can shine.

Pusanga, then, along with the oils of attraction, are completely natural forces that allow us to free our minds from the conditioning that says *We cannot have, We are not worthy, We should feel guilt at having this,* so we can find our spirit and achieve balance once again. The mystery of pusanga is the mystery of ourselves, and the magic it amplifies is our own.

To understand this more fully, try the exercises below.

Why Can't I Have What I Want?

This is a simple exercise, though you might find it revealing. Close your eyes and relax . . . and now, silently or out loud, ask yourself, *Why can't I have what I want?*

Watch for any images that come up and listen for any words that pop into your mind. Whose words are they, whose voice is speaking, and what is it saying to you? Examine any information you receive, with the object of getting very clear on what you are being told to believe. The messages you received in the past may have been true for the messenger, but may not be true for you. When you examine them in the light of your own day, you can choose which of these beliefs are useful to you, and which are cluttering up your life and keeping you from having what you want.

While doing this exercise, one workshop participant, James, recalled an event from his early life when he was taken to visit his grandfather. It was the first and only time he met this man, and he'd forgotten all about it, but now he saw it all clearly again: the room, the look of his grandfather, the clothes he wore; and he remembered sitting on the old man's knee before an open fire.

Grandfather's words of advice to his grandson were these: "Be careful. It's a dangerous world and there are plenty of people out to get you, so don't make waves, don't raise your head too high, and don't ask for too much. That way you stay out of trouble."

Consciously, James had forgotten all about this—but somewhere within him he had stored this advice and, *unconsciously*, he had been acting on it for most of his life. The fact that his grandfather had died soon afterward added force to the words: perhaps the old man was right and "they" *had* "got him" because he had asked for too much or revealed a great secret to his grandson.

James's grandfather had raised thirteen children. He was born and died working class; his job was in a factory, working with asbestos, and asbestos poisoning finally killed him. Given his circumstances, everything he had told James probably had some basis in truth: it *was* a dangerous world. But that was also *his* truth, *not James's*.

When James realized this and heard his grandfather's voice again, he made his conditioning conscious and could then let it go.

Whose voice do you hear? And what are you being told to believe?

🪶 Making Pusanga

Making pusanga begins with a plant walk. This is like the walks of attention you have taken earlier, in the sense that you relax and center yourself so that your walk is partly conducted in dream space, allowing you to slow down to the pace of nature. On this walk, however, you are not intending to hear the call of a particular plant. Your intention is to actively stalk the plants you will use in your mixture.

The doctrine of signatures is your guide. Pusanga plants all have signature characteristics. Those for love may have names that are meaningful, such as *passionflower* or *honeysuckle*. Their colors will be bright and attractive. The way they grow may be significant (ivy, for example, winds itself around other plants so the two intertwine and are drawn close together). Their archetypal qualities may also call to you. (Rose, for example, is nowadays practically synonymous with love, but other plants may have qualities you'd also like in your relationship or from your partner—fern is one of the oldest plants on the planet and signifies endurance and timelessness; oak is a symbol of wisdom; and so on.) Where the plants grow might also have meaning (two plants standing together in sunlight within an otherwise darkened forest may signify a bright future, for example), and so on. Look for plants that mean something to you and for the situation or desire you bring to your walk.

When you locate each plant, spend a little time with it, explaining your need and asking if the plant will offer a little of itself to you. (You don't need to take the whole plant; a single leaf, a flower, or a piece of bark will do, as this contains the energy of the whole. Try to avoid taking roots if you can.)

If the answer you get is *Yes,* take a piece, and offer your thanks to the plant, and perhaps a gift of your own. In Haiti, coins are often left at the base of a tree whose leaves are picked; in North America, corn or tobacco may be a suitable gift. Be guided by your intuition.

What if the answer is *No?* Simply move on to the next plant. But also be aware that a *No* can be useful information too. Remember that nature has an affinity for human beings and wants to help you, as we saw in chapter 1; it will not deny you what is in your best interests. Remember also that pusanga is about *you.* If you receive a *No* from nature, therefore, it may serve to ask what, within you, wants to deny yourself this plant and the qualities it offers. Plants are capable of ESP, as we have seen, and it may be that nature is giving you the very thing you are really asking for: a No. If you feel this is the case, it will be useful to return to the preceding exercise before you continue your walk.

When you have the plants you want, take them home and put them in a clear bottle. If you intend to use the pusanga over a matter of days, you can fill the bottle with water. Water from "power places" is best—holy water from churches or water from a place of spiritual power, such as the Chalice Well at Glastonbury—but you can also use pure spring or mineral water. If you want to keep the pusanga for a while, it is better to use alcohol instead of water as this will preserve the plants. You can also add agua florida or perfumes to the mix, and again, consider the name of the perfume you choose. Names like Cachet suggest wealth and status, for example, while White Satin suggests romance, and Chanel is associated with love and luxury. In some Vodou mixes for love, Reve D'Or (golden dream) is specified for obvious reasons. Choose your perfume according to what you want to attract.

When you have done all this, sing your plant song into the bottle to create a deeper connection between you and the plants, then add your intention by blowing three blasts of tobacco smoke into the pusanga bottle. *Salud* (health), *dinero* (wealth), *y amor* (love) are the traditional prayer words used to empower a pusanga, but you can, of course, ask for anything you want.

As the final ritual before you use the pusanga, go out into a forest or park and find a tree that you feel drawn to in some way. Tip three drops of your pusanga at the foot of this tree, and with each drop say out loud the things you want to attract. When this is done you can use the pusanga on yourself and in ritual magic.

Pusanga is best stored in a fridge and will last longer that way. The plant material it contains will degrade over time, of course, but this is irrelevant to its magical effects and the mixture can still be used.

There is one other thing that is important when you use the pusanga: *you must believe in its ability to work for you "without an atom of doubt."* You have a *right* to the things you have asked for and you *can* have anything you want. Nature is your ally in this.

6

❧

FLORAL BATHS:
BATHING IN NATURE'S
RICHES

*Nature, whose sweet rains fall on just and unjust
alike . . . she will cleanse me in great waters, and with
bitter herbs make me whole.*

OSCAR WILDE

In Celtic myth, the gateway to the otherworld was provided by natural forces, and often by plant magic. One day, while walking the hills, Cormac, grandson of Conn, met a mysterious grey-haired warrior who gave him a silver branch from a sacred apple tree. The warrior vanished as soon as he'd given the gift. When Cormac shook the branch—as so often happens in these tales—a strange mist enveloped him and he was lulled to sleep by the music of the leaves.

He awoke to find himself in the Land of Promise, where only truth is known. Great bronze palaces and feather-thatched houses of silver stood all around him, and at the center of this, a glistening fountain from which flowed five shining streams. Nine hazel trees grew above the fountain, dropping their seeds into the water, from which fairy folk would drink.

Cormac was taken to the fountain by maidens, who bathed him in the waters and gave him a magical gift: a cup of truth, which would shatter if three lies were spoken in its presence, so that Cormac could discern honesty from falsehood. He returned home with this gift from the Land of Promise and became a great leader of men.

The "wisdom bath" that is used to anoint Cormac is an initiation into truth. The waters cascade from the Fountain of Knowledge, and the streams that flow from it are the five senses through which the world is experienced and wisdom obtained. The hazelnuts that dance in the waters are, as we saw earlier, also symbolic of wisdom.

This, then, is a cleansing bath, one with which many shamans would be familiar. It opens the doors to true perception, washing away illusion and self-doubt, and replacing them with the clarity and self-knowledge needed by a leader. Once again, the plants provide the gift.

In *Welsh Herbal Medicine*, David Hoffmann writes that Celtic "Druidic medical therapeutics is an interesting combination of mystical and herbally rational techniques. For internal and lingering complaints they mainly used the cold bath . . . with the administration of herbs." He continues:

> Great use was made of water from certain wells, due to their specific mineral and spiritual properties. . . . The Druids devoted considerable effort to study the medicinal properties of plants, believing some herbs to be endowed with magical virtues. [Such herbs and baths were used for a variety of purposes, including] to anoint people, to prevent fevers, to procure friendship and to obtain all that the heart desires.[1]

ALL THAT THE HEART DESIRES

Herb baths such as these are not just confined to the Celtic healing arts. They are used in many cultures to purify, to give deeper knowledge of self, to free the soul of impurities and limitations, and to bestow spiritual gifts and physical healing. Every bath has the same ultimate purpose: to bring the bather into balance by washing away fear, anger, trauma, grief, and uncertainty, and giving inner peace and a deeper connection to nature. These baths cleanse the soul and are known in many traditions for their power to do so.

In Malaysia, for example, the *mandi bunga* bathing ritual is carried out to invoke the powers of nature so that a person may achieve whatever he or she most desires. The ritual is undertaken for a variety of reasons—to find love, to remove *buang suey* (bad luck or evil magic), to heal body and soul, and even for seemingly mundane and "non-spiritual" matters, such as good fortune in gambling and games of chance.

The ritual is facilitated by a *bomoh* (shaman) who prepares a bathing mixture of plants, flowers, nuts, leaves, fruits, and spices such as *limau purut* (kaffir lime) and *akar sintok* (blume) that will be uniquely formulated to the patient's needs. Other materials may be added, such as *bedak sejuk* (powdered rice grains) before the blend is mixed with water and the patient is bathed.

In a ceremony to capture a husband, for example, a woman will be bathed in waters containing fruits, leaves, and seven types of flowers, while the bomoh burns a candle wrapped in coconut leaves, all the time chanting and sprinkling grains of rice around his patient. They say that this brings marriage within the year.

In nearby Indonesia, there are almost 7,000 recorded varieties of plants (by contrast, there are just 1,600 in the United Kingdom), and in this rich environment many of the herbal secrets once guarded by priests and princes have now become so commonplace that they are standard treatments in salons and spas. *Mandi luhur* (coating the skin) is one traditional therapy that has been practiced since the seventeenth century. A paste is made of spring water and herbs, including sandalwood, turmeric, and jasmine; this is massaged into the body and left to dry. The result is beautiful, healthy skin, as well as inner peace.

Another ancient recipe, used to rebalance the body and as a cure for rheumatics and colds, is a paste of sandalwood, coriander, turmeric, cinnamon, ginger, cloves, and nutmeg mixed with water and scented oil. It is no accident that many of these ingredients are also used in cooking, since the intention of this medicine is to "cook" the body by producing a deep heat. The patient is revitalized by this, and when the paste is removed, it draws the body toxins out.

Hot wraps made from mud and herbs are also applied to the skin so the body can sweat out its physical and spiritual impurities. One of these has been used by women for centuries to regain their figures after childbirth. Every day for forty days following the birth, a mixture of lime juice, betel leaf, eucalyptus, and powdered coral is blended to a paste and applied to the abdomen. A sheet is then wrapped tightly around the stomach and the plants are left to work their magic.

Flower and water baths are also used extensively in plant spirit therapies. In many shamanic cultures, the healer uses cold water to stimulate the body and soul; but in Indonesia, warm water is more common, the aim being to relax the patient so his energies can flow unrestricted by inner tensions. The flowers most frequently used in these baths are magnolia, bougainvillea, hibiscus, rose, gardenia, and jasmine, which envelop the patient in a powerful healing scent, while their vibrant colors lift the spirit and reflect the beauty of nature to remind the patient of his own inner beauty and strength.

BATHS THAT CURE NIGHTMARES

Maya shamans also use baths like these to restore the harmony of the soul. One such bath is to cure children of nightmares, which the shamans see as a spirit that is dark and shaped like a human embryo. This spirit enters the body when soul loss (see chapter 4) has occurred as a result of shock or grief. Because of their tender years and greater vulnerability, children are especially open to spiritual illnesses, which, along with fear, envy, and depression, can cause the loss of *chuílel*—spirit, or life force.

An American mother, Nadine Epstein, recounts her use of such a bath to help her six-year-old son, Noah, overcome his recurrent nightmares. In an article that appeared in *Mothering* magazine, Epstein recalls the desperation that led her to try this method after all else had failed and Noah's frightening and violent dreams continued.

One day, when Noah seemed particularly burdened, we had an idea. Why not give Noah a spiritual bath? The Maya regularly give them to children, even infants.

> So late that afternoon, when the light was beginning to turn to lovely slanting gold, [my friend] Rosita and I took Noah out on a ritual plant-gathering foray into our urban neighborhood. Baskets in hand, we wandered along sidewalks and alleys searching for healing plants that could be used for spiritual bathing. . . . When we came upon a plant, we carefully plucked it, saying the Maya prayer to thank the plant and asking the plant spirit to help heal the heart and soul of Noah. . . .
>
> We then filled a bathtub with water, and let the infusion of aromatic plants loose.* They floated over the surface of the water, creating an intricate, colorful pattern. . . . Once in, [Noah] immediately became absorbed in propelling the leaves and petals in spiral patterns around the tub. The hollyhocks felt cool and soft, and he rubbed them against the skin of his knees. As he played, we said more prayers and burned copal incense. . . .
>
> When Noah finally did emerge from the tub, the difference was amazing. He had lost that feeling of heaviness that he often carried around with him. I had not seen such a deep transformation . . . Noah didn't have a nightmare that night or any night in the next few weeks.[2]

Epstein gives the recipes for a few of the children's baths she has studied. A calming bath for infants includes marigolds, basil, St. John's wort, rosemary, and rose petals, which are squeezed to release their essence and added to warm water. A bath for skin infections uses the bitter leaves of motherwort, wormwood, and/or dandelion. These are boiled in water before the child is bathed and the infected area cleansed with the mixture.

---

*In this bath, for removing the spirit of nightmare, the key ingredients are rose petals, hollyhock, rosemary, and marigold.

BATHS THAT MAKE RAIN

Healing baths are not always restricted to personal healing. Because of the energetic connection we share with the whole of nature, they can also be used in communal healings to create transpersonal effects, such as the examples Frazer gives in *The Golden Bough*, where bathing is used to make rain.[3]

In India, when rain is needed and crops will not grow, the villagers dress a young boy in leaves, and he becomes the Rain King who controls the waters of heaven. Every villager takes part in this ritual, each of them splashing the Rain King with water as he goes from door to door, and each offering a gift to "buy rain."

A similar ritual takes place in Russia, where women bathe publicly on the day of St. John the Baptist, using water that has been dipped with a fetish of grass, leaves, and herbs to represent the saint. The fetish *becomes* the saint, and once the women have bathed in his waters and taken on his essence in this way, they, too, have the power to call rain.

In New Caledonia, a somewhat different ritual is followed, when rainmakers bathe a corpse that is suspended over taro leaves in a cave. The soul of the ancestor transmutes the water of the bath and releases it to the villagers as rainfall.

HOW DO PLANT BATHS WORK?

There are three principles behind the effectiveness of baths such as these—baths that revitalize body and soul, as well as those that produce more transcendental outcomes like rainmaking. Readers will recognize all three principles from earlier chapters and from the work they have already done to understand the nature of plant spirits. The first principle, of course, is that plants are composed of sentient energies and have a spirit that is inclined toward human beings.

The second principle is that human beings are composed of this same energy. Nobel-nominated physicist Michio Kaku has remarked, for example, that the solid parts of the human body, if compressed, would only occupy a space of a few inches. The rest is energy, and this is what holds our atoms together.

If you find that hard to believe—that our "solid" bodies are really just inches in size—you are not alone. Kaku himself remarks that, for scientists, too, "the universe is not only stranger than we imagine, it is stranger than we *can* imagine."[4] That we are beings of energy, however, is not in dispute.*

The third principle is that this is also true of everything in the universe: it is all made of energy at its most fundamental level, and, as we know from the science we learned in school, energy cannot be created or destroyed, it can only change its form. It is cooperative and can be shared, in other words.

Shamans, of course, would not put it this way. They would simply say that everything is spirit and, if we are sensitive to this, it is quite possible for the spirit of a plant to enter and influence us, changing our energy and creating new possibilities for healing. And they would say that it is equally possible for us to communicate with other energy beings, such as rain spirits.

Let's consider these three principles in more detail.

PLANTS AS INTELLIGENT ENERGY

In shamanic cultures, plants are regarded as "the transformers." They take energy from one place and one form (the sun, soil, and rain) and transform it into another form: food. The most important of these original energy forms is sunlight.

This, in itself, is remarkable: *plants eat light.* Just as remarkable is what happens after their meal is over. *They excrete oxygen.* Plants, just by eating, achieve the alchemically spectacular. They take earth, fire, and water and produce air. And, through this, they give us life.

When plants die, their energy remains, and it is *still* useful to us. The plant, and the light in its leaves, will eventually mulch down to

---

*For a discussion of how this relates to shamanism, see Ross Heaven, *The Journey to You* (Bantam Press, 2001), and *Vodou Shaman* (Inner Traditions, 2003).

form "fossil fuel" and bequeath its sun-energy to us. Whenever we burn oil or coal, we liberate an energy that was stored in the plants a million years ago. The same is true, of course, whenever we eat a plant: we complete the circle, fire to fire, and we too become consumers of sunlight.

Plants *are* energy, but they are also intelligence. Jeremy Narby makes a study of plant sentience in his book, *Intelligence in Nature,* and concludes that they are, in fact, not so different from us:

> Science now indicates that plants, like animals and humans, can learn about the world around them and use cellular mechanisms similar to those we rely on. Plants learn, remember, and decide. . . .
>
> Plant cells relay information to one another using signals such as charged calcium atoms. Our neurons do the same. Plant cells also have their own particular signals, which tend to be relatively large and complicated proteins and RNA transcripts. These molecules swim around the plant providing information from cell to cell. Individual plant cells also appear to have a capacity to know. . . . Scientists now confirm what shamans have long said about the nature of nature.[5]

The fact that plant signals (proteins and RNA transcripts) tend to be large and complex, in contrast to the small molecules of human brain signals, is also significant, since "large molecules can handle large amounts of information, which means there is room for enormous complexity in plant communication."

Part of the scientific prejudice against the notion of plant intelligence is based simply on the fact that they do not move. Old-school scientists have concluded that there is therefore no evidence for intelligence, because only deliberate movement suggests the application of thought and decision making. But this is just silly. Plenty of people sit still and think.

And, in fact, if we look beyond our prejudice, we find that plants *do* move; they just work on a much slower timescale, so we tend to miss their movements. Narby gives the example of the tropical stilt palm, which "walks" toward light by growing new roots on its sunny side and allowing those in the shade to die. "By doing this over several months, the stilt palm actually changes place."[6]

Another example is ground ivy, a vine that creeps across the land and only puts down roots when it finds the correct balance of soil nutrients and light. By doing this, it pulls itself to new locations by skipping the places it *knows* are unsuitable. (If we look at the migration of human beings, they, too, with varying degrees of rational thought, move to new locations that will sustain life, in a similar way to ground ivy and stilt palms. Perhaps this is the basis for the shaman's observation that the Creator—that is, the creative force, nature—ensures that the plants we need for food and healing always grow where we live. Plants and people move toward one another, drawn by the same attractive force: an understanding that this new environment is most suitable to their mutual needs.)

And, in fact, anyone who has seen the remarkable award-winning BBC television series, *The Private Life of Plants*, by David Attenborough, can observe for themselves that plants move. In one episode, time-lapse photography was used to record the movement and growth of rainforest plants. With filming time speeded up, all the plants in the rainforest are shown to be in a constant flux of movement, striving to reach the sunlight at the tree canopy and the water under the soil. Vines swing to and fro, plant tendrils reach out to other plants to gain purchase in their quest for sunlight. The forest is a mass of movement; it is just that we do not perceive it because our senses don't operate on "plant time."

All indigenous peoples appreciate these plant qualities. In her book, *Magic From Brazil*, Morwyn reflects that "During germination and growth, plants absorb and store immense energies from the earth and sky. When a person ingests [a plant] the energy is freed and circulated throughout the body and into the aura. The herbal energy both adds something of its own nature and helps release the patient's own pent-up energy to stimulate self-healing."[7]

Scientists have recorded this "herbal energy" many times in plants, most famously through the use of Kirlian photography, a method invented in Russia by Semyon and Valentina Kirlian and now used in research laboratories worldwide. The principle behind Kirlian photography is that all living beings emit radiation in the form of light, electromagnetic frequencies, and heat in direct relationship to their internal states. Kirlian photography captures and records these emissions by introducing a high-frequency, high-voltage, ultra-low current to the subject being photographed, which amplifies and makes visible the energy it contains. Among the successes of this method are its abilities, in controlled experiments, to predict the survival, growth rates, and general health of various seeds and plants purely by reference to the energetic "aura" of the plant that is captured on film.

Some of the more interesting of these experiments are inquiries into the "phantom leaf" phenomenon. This is where a small portion of a leaf is removed and the plant is then photographed. A ghostly image appears of the missing leaf section, exactly where it would be if it was still attached. Kirlian researchers see this as proof of a nonmaterial, "bioplasmic body" within the leaf, an energy that, though nonmaterial, is essential to life.*

Because of their energetic powers, in Candomble (the Brazilian form of Vodou about which Morwyn writes), "Plant materials enter into every aspect of ritual. . . . Each herb is believed to possess an etheric force easily capable of absorption by the skin. Every botanical crystallizes a particular virtue such as fertility, peace, vigor, protection, longevity, courage, happiness, good fortune, and glory, and may also drive away illness, negativity, misery, and noxious fluids."[8]

Specific baths, called *abo*, are used to purify initiates for priesthood. Others, called *banhos de descarga* (discharge baths), neutralize negative energies that have attached themselves to a person. The *amaci* is a head bath that strengthens the connection between a person and their *Orixa*,

---

*There is also a correspondence to the "phantom limb" sensation reported by people who have lost arms or legs in accidents but still experience the limb as present, when, clearly, there should be no sensation at all unless they are experiencing something beyond the physical. For more on this, see Shelia Ostrander and Lynn Schroeder, *Psychic Discoveries Behind the Iron Curtain*, Prentice Hall, 1984.

or guiding spirit, as well as shielding them from negative influences. And there are many more.

PEOPLE AS INTELLIGENT ENERGY

Instead of using the word *soul,* some shamans talk of people having an "energy body." This equates with the bioplasmic body discovered by Kirlian scientists in plants and human beings. The energy body can be imagined as a luminous egg that encircles the physical body at a distance of about an arm's length in each direction, though it is slightly closer at the head and feet.

The term *luminous egg* as a description of this energy has passed into contemporary usage from the work of Carlos Castaneda. As a consequence, it has been somewhat dismissed by those who suspect Castaneda of fabricating his encounters with the Yaqui shaman, don Juan, who supposedly invented the term. But in fact, this vision of the energy body predates Castaneda by centuries. In the European Romani traditions, the soul was also described as egglike. In Haitian Vodou and Cuban Santeria, eggs represent the soul, and even the Catholic Church likens the soul to an egg.

The Andean shaman Doris Rivera Lenz (interviewed in chapter 1) says of eggs that "[They] are the union of the masculine and the feminine. We should recognize that this union is supremely sacred. We are the product of an egg too. So the egg is the total energy of the mother's and the father's cells."

Quite consistently, then, across many cultures, the energy body is regarded as egg shaped. And it is this that shamans see when they enter their healing trances.

The energy body is sometimes further divided into four bands of light, representing different aspects of the self. Typically, from the outer band inward, these are the spiritual, emotional, mental, and physical "bodies" that make up the soul.

The spiritual is the furthest from the physical body, beginning at a distance of about an arm's length and stretching into infinity. This is what gives us our connection to the energy universe as a whole. It is through this that ESP and other unexplained phenomena operate and it is this that facilitates our communication with other intelligences, such as plants. The emotional self is slightly closer, in a band that begins about twelve inches from the body. And the mental self is the space between the emotional and the physical.

All illnesses stem from the unseen world and affect our spirits first. If we are sensitive, we may detect an intrusive energy or "outside force" as soon as it connects with this band of light and sense that something is wrong, even if we can't say what. This is the feeling of walking into a room and being uncomfortable for no apparent reason, or meeting a new person and knowing they are not for us. Many people have these experiences but are less aware when it comes to matters of health, so the presence of intrusive energies may go undetected until they are established in our fields.

As it begins its migration toward the physical body, the spirit-illness will, however, become more noticeable. It is likely to register first with the emotions, since these are our most sensitive organs of awareness. But it is only when the illness enters the mental sphere that we might be able to sense and name the problem. Finally, what began as a subtle spirit awareness becomes a physical reality.

In conducting her healings, the plant shaman therefore concerns herself with the spiritual point of origin. If she can disperse the intrusive energy before it becomes solid and ingrained in the patient, every part of the energy body, from the spiritual through to the physical, will be healed. This is often the purpose of the sacred bath: to remove unhelpful energy and replace it with a force that is helpful and good.

Such healing addresses not the *symptoms* of illness within the physical body, as Western medicine is apt to do, but the *cause* of the illness itself, on the basis that, since all diseases are energetic, they can be cured in the same way: by changing the energy of the patient. Shamanically speaking, there are a limited number of reasons why illness occurs, and these usually have to do with a lack of balance or a spiritual disconnection within the patient.

In Vodou, for example, maintaining a good relationship with God, the Lwa (guardian spirits), and the Ghedes (the sacred dead and the ancestors) ensures the health and power of the individual. It is only when this connection is broken that illness is possible.

How we might express this in Western psychological terms is that when we have faith, a reason to believe, and a purpose in life, when we have family support and embrace our lineage, we also have a strong self-identity and sense of worth, so we automatically feel healthy, fit, and "in our power." A number of scientific studies have shown, for example, that people recover more quickly from hospital operations when they have faith as well as family and friends who care for them.*

When people are alone in the world, however, their self-esteem drops and depression sets in, so that health suffers and they can lose their will to live. The psychiatrist Viktor Frankl wrote of this in his book *Man's Search For Meaning*. For Frankl, the isolation and the lack of purpose and connection in modern life were the chief causes of mental, emotional, and in many cases, physical problems in the clients he saw. Hope—something to believe in—and connection with others were the means of a cure, more important even than the psychotherapy itself.[9]

Houngan[†] Max Beauvoir writes the following:

> Many illnesses are considered to result from the individual's own inappropriate social behavior. For instance, often cases [in Haiti] pertain to inheritance disputes in which the individual is perceived to have wrongly appropriated family common land. Or, a lack of respect, of courage, and/or of generosity, which may in themselves constitute the so-called weakness of character, is known to reflect on the individual's own health. In these situations, it is said that the person is persecuted by the spirit of a dead person (mò), sent upon him by his neighbors or family.
>
> Here, it is necessary to have the patient redeem him or her-

---

*See Ross Heaven, *Spirit in the City*, for further discussion of this.
†A houngan is a shaman priest and healer in the Vodou tradition.

self. Pilgrimages, charity visits to hospitals and food contributions to prisoners and to the paupers, coffee for the ancestors, masses for the deceased, and other such actions, constitute elements susceptible to bring about a remedy.

Furthermore, the invading "mò" is often chased by resolute means, such as a good flogging with pigeon pea stems *(Cystisus cajan, L.)*, which is considered to be one of the most radical and efficient means of expulsion.[10]

Individual "weakness of character" might in some ways be regarded as a social rather than a wholly personal disease, especially in societies that invite competition, with social approval for those who "make it" and punishments (social disapproval and outcasting) for those who don't. In such cultures people often feel scared, hopeless, lost, and alone. Surveys tell us, for example, that many people in the West no longer trust politicians, big business, or even food suppliers, whom they believe to be interested only in profits and their own well-being. They see themselves as leaderless and trapped, isolated and betrayed, and this leads to frustration, self-doubt, and anxiety. Tensions like these can in turn lead to physical illnesses, as well as a weakening of the soul, which inclines some people toward immorality.

One way that the Vodou tradition deals with this disconnection is by using bathing rituals known as *lave tet* (head washing), and in Santeria as *rogacion*. The purpose of head washing is to "open the head" of the patient and reconnect him to God, the Lwa, and the ancestors. In psychological terms, we might say these baths relieve tension by loosening the patient's attachments to competitiveness and helping him relax into a deeper meaning and a healthier connection to the universe. The patient no longer obsesses about the material world because he sees something beyond it and understands that he is as loved by God as anyone else.

The head washing concludes with ritual songs, all of which, in one way or another, reinforce this sense of equality before God. *"Nous tout se yon-O! divan Bondye"* is one such song. Translated, it declares that "We are all one before one God," Bondye (from the French *bon dieu*) being the Haitian name for the supreme energy of the universe.

Beauvoir writes of bathing rituals like these, that "Immediate results emerge from the energy charge which flows from these herbs and the energy used to macerate them. Singing and dancing, the common forms of Vodoun expression, are themselves considered prophylaxis, promoting a healthy distribution of the Dan [spiritual energy or life force] throughout the body."[11]

THE UNIVERSE AS INTELLIGENT ENERGY

String theory is one of the more interesting developments in recent physics. It was developed as an attempt to overcome the historical failing of mechanical science to explain the universe and to make sense of the contradictions and anomalies we see in the world. As Cambridge University's online publication *Cambridge Relativity* puts it: "To take into account the different interactions observed in Nature one has to provide particles with more degrees of freedom than only their position and velocity."[12] To effectively describe the world, in other words, one has to allow that it has more dimensions than the four we can observe: length, breadth, and depth (in space) and the fourth, time.* *Cambridge Relativity* continues:

> Theories were built which describe with great success three of the four known interactions in Nature: Electromagnetism, and the Strong and Weak nuclear forces . . . unfortunately the fourth interaction, gravity, beautifully described by Einstein's General Relativity (GR), does not seem to fit into this scheme. Whenever one tries to apply the rules . . . one gets results which make no sense. For instance, the force between two gravitons (the particles that mediate gravitational interactions), becomes infinite and we do not know how to get rid of these infinities to get physically sensible results.[13]

---

*Even these four dimensions may be questionable as they stand. Time, for example, is a human construct, because all we really observe in nature is change. *Time,* like *aging,* is a relative term that we apply to this change.

There are two points to note here. The first is that modern science also regards the universe as fundamentally composed of energy. This is what these "four known interactions in Nature" (electromagnetism, strong and weak nuclear forces, and gravity) describe. The existence of physical matter and physical effects (and of the universe itself) therefore depends on energy.

The other point of note is that scientists still, and by their own admission, do not understand how the universe works, since one of the basic energies they are attempting to deal with (gravity) does not fit their own rules. And yet the world still works. This is one justification for Michio Kaku's statement that the universe is stranger than we can imagine.

For some scientists, the conclusion (most often spoken in hushed tones and whispers in the halls of academia) is that there is an intelligence to this energy of the universe, an intelligence that we don't understand and cannot measure, but which is greater than our own. The physicist Dr. Fred Alan Wolf, for example, in *The Dreaming Universe* talks of his realizations about the world, that no "forces had created this blindly, nor had mechanics created it nor had blind nature created it. A clearly organized, intelligent, feeling, sensing, like-myself, anthropomorphic being had created it. In that sense I felt the presence of God."[14]

Even scientists who overtly deny this intelligence must tacitly accept it, if they are working on string theory at all, for the very name of this branch of science comes from the fact that the operations of the universe, within this scientific model, are conceived as vibrations on the string of a vast musical instrument. And every instrument must be played by someone —or something. Music describes the universe not only more poetically, but, these days, more accurately than mathematics.

For our scientists as much as our shamans, the universe is energy and it is intelligent and self-aware. Human beings, plants, and the entire world order are composed of this energy. That is why something as simple as a floral bath can work: it works through the energy that we share. The spirit of the plants puts us in touch with the greater energy of the universe and the intelligence pervading it. Or, as the shamans would say: we are healed through our reconnection to the divine spirit of nature.

BATHS IN THE AMAZON:
AN INTERVIEW WITH A SHAMAN

In Peru, floral baths are known as *banjos florales* (flower baths) and are a staple of shamanic healing from the high Andes to the Amazon basin. People use them to wash away unhelpful spirits in order to remove blockages, so that the energy of the universe can flood in and correct the imbalance. Artidoro Aro Cardenas, describes the process in Peru.

How are these baths taken?

AAC: *The bath is most often taken on the morning after ayahuasca ceremonies so that the body is modified to accept the new information of the visions. But this is not always true. Sometimes baths are taken before the ceremony to open the person up, and sometimes they are taken by themselves, as a healing.*

A tub is filled with water and to this is added the plants that the patient most needs, like mocura and ajo sacha, some of the most powerful doctors. Agua florida or agua de colpas may also be added.

The person must approach in a sacred manner, in prayer that his needs will be met, and with the intention that they will. The shaman then pours the water over his head and lets it run down his body, also blowing with smoke to purify him, or with perfume so he will flourish.

Sometimes the person turns as this is happening—first to the left [in a circle, counterclockwise], then to the right [clockwise]. The first turn is to get rid of negativity; the second to draw in positivity.

*The bath takes place on the bank of a river so the energy that is removed will find its way to the sea.**

What plants are used in baths?

AAC: *Floral baths do not contain large numbers of plants. Specific plants or flowers are chosen instead according to the patient's ailment.*

I begin by cooking up good-smelling plants from the forest, and to

---

*That is, be taken away completely.

that essence I add a little alcohol and a little agua florida. Then I get flowers and mash them and add that juice to the mixture and put it into bottles. When I do this, I diet and refrain from eating salt, etcetera. You can either have a one-off floral bath or you can have a series of them for a deeper and more thorough effect.

A common reason for people to want to take floral baths is that something is not going well for them—like, for example, they can't get work or they are having bad luck. First I give them a cleansing bath to take away the saladera [bad luck] which shows up as salt on their skins. In that bath I put ajo sacha, mishquipanga, ruda, and romero [rosemary]. Then the floral bath follows to give the things the client wants: luck, work, etcetera.

Can you give examples of other baths and what they are used for?

AAC: For changing luck, mocura is used, and the patient will find that after a couple of weeks, things have changed. For example you may find the job you were looking for, or where your life felt stuck or turbulent there is some momentum; things start to shift. Mocura is also used for clearing negative thoughts and feelings sent to you by others.

For cleansing the spirit, the dark red leaves of pinon colorado are used to undo sorcery and harm. This plant is also used in steam baths and when this is done you can actually see the phlegm, which is the bad magic, appear on the patient's skin as it comes out of the body.

For flourishing or blossoming, bano de florecimiento plants are used. These help us to connect with and draw upon the strength and courage within ourselves, to overcome obstacles, and to lead a purposeful and productive life in accordance with our soul's intention. The mixture for this bath is agua de colpa water from a place in the forest where pure rainwater collects. Often hunters drink this water as well to attract the animals. To this is added albacca, which is a plant used widely in Peru for its strong, sweet perfume. It is used instead of an aerosol spray to freshen a house and is also placed on corpses during funerals. From a floral bath perspective, it attracts lots of friends and positive outcomes. It is also used medicinally for gastritis, appendix, or gall bladder problems, in which case you can take it as a tea. Menta [mint] is also added to freshen and revitalize the *bather. Menta is also good for calming the nerves and releasing worries and preoccupations. When the person bathes, all of these plant qualities are absorbed by the skin and the spirit.*

Although the exact ingredients may not be available in the West (with the probable exception of mint), it is possible to work with plants that have similar qualities to the ones Arevalo describes.

If you would like to try the luck bath for yourself, for example, in place of agua de colpa, you can collect fresh rainwater that has gathered in a forest, or the morning dew on grass and leaves, and use it as the base for this bath. In place of albacca, use other fragrant flowers or herbs—rose petals, for example, or saffron that has been crushed to release its fragrance. (Also see appendix 2 for other alternatives.)

Again, it is a positive attitude and a clear intent that is most important. The plants will then help you to draw in benign and benevolent energies.

BATHS IN HAITI:
AN INTERVIEW WITH A LEAF DOCTOR

The ritual bathing process is similar in Haiti. Here, the bather approaches the water barrel carrying a white candle, which he must keep alight at all times. No doubt this is to focus his mind on a prayerful intent for his healing, although candles are also a way to attract the attention of spirits.

The candle is handed to an assistant and the patient kneels before the barrel to receive his first wash from the shaman or medsen fey. He then stands and, in the same way as in Peru, turns to the left and the right for subsequent washes. In Haitian magic, counterclockwise is the direction of decrease and removal, and clockwise is the direction of increase and drawing in.

The number of washes he receives and the plants contained in the waters depend on which Lwa has, through the prayers of the shaman, given power to the bath. "This depends on the patient's needs," says Loulou Prince.

What plants are used?

LP: The plants that we use are secret, but I will tell you, for example, that Papa Ogoun [the Lwa] brings power and he likes spices and hot peppers, so we add these things to call him into the bath. It is never just plants you receive; the plants are the Lwa.

They [plants] also contain Gran Bwa, the spirit of all the leaves in the forest. When I bathe a person and rub him with these leaves, I am all the time praying to the Lwa to bless and protect him, and he takes in the Lwa themselves; he is not just rubbed with plants!

What rituals are involved in the bathing?

LP: The ritual begins long before the bath itself. I must go out to the woods and collect the leaves in a proper manner, and pay the trees. (See a photograph of a houngan leading a leaf walk on page 4 of the color insert.) Then there may be a pile fey,* a ritual where the leaves are ground to a powder for magical use.

The mortar used in a pile fey is a large hollow tree stump and the pestles, also made from the trunk of a small tree or a thick branch, are so large that two men are needed to pound the leaves. Often, the Lwa themselves possess the men to help them with the work and, to encourage this, there is dancing, drumming, and singing during the ceremony.†

The leaves are then put to a variety of uses and become the material for pakets, wangas [magical charms, e.g., for luck and love] and, of course, the baths. (See the photograph of wanga bottles decorated with serpents in the color insert.) Only when all these appeals to the Lwa are made and our respects have been paid should the patient be bathed, or it will not work.

---

*Literally, "breaking leaves."

†Several of these songs again make a connection between plants and serpents, a relationship that also exists in the Amazon, where ayahuasca is believed to be born from a snake. This is true even though ayahuasca is unknown and unused in Haiti. One medicine song, for example, has the following words: "Fey, koulev o, fey! Fey o, koulev, fey! M pral benyen ti moun yo!" This translates as: "Leaves, serpent, oh, leaves! Leaves, oh serpent, leaves! I am going to bathe the children!"

There is a fine balance [to this healing]; if the patient takes on too much power, it is said he may become loupgarou,* so he is wild and out of control. Balance is the key to healing.

We explained to Loulou that in Jamaica, a nearby Caribbean island, ritual baths take place in "Balm Yards," dedicated spaces where the doctor and priest bathe a patient. According to Zora Neale Hurston, in *Tell My Horse*: "Sometimes [the healer] diagnoses a case as a natural ailment and a bath or series of baths in infusions of secret plants is prescribed. More often the diagnosis is that the patient has been 'hurt' by a duppy [a spirit], and the bath is given to drive the spirit off. The Balm Yard with a reputation is never lacking for business."[15]

Where do your bathing rituals take place? Do you also use a dedicated sacred space like the balm yard?

LP: *In Haiti, the bath is often taken in the hounfor [the Vodou temple], and there is a special area I use for this. I suppose this is like a "balm yard," although it is less formal than that.*

Baths may also take place outside the hounfor though, at springs and rivers where the Lwa live. Those for La Siren, the mermaid spirit, may be received in ceremonies conducted at the ocean.

There are also pilgrimage sites for luck baths, such as those at Bassin Saint Jacques in the Plaine du Nord, where the bathers wallow in mud which is charged with special energy.† It is common for possessions to occur at this time as the Lwa enter the body of the bather to purify him from the inside.

Another pilgrimage site is Saut d'Eau, which has waterfalls surrounded by huge trees that are the homes of the spirits. Through the leaves and roots of these trees, the spirits charge the waters with their power.

---

*Literally, a werewolf.

†This is a form of "rebirthing," in fact; an ultimate healing or blessing, where the bather "disappears" in the mud and is then reborn as spiritually and physically clean when the mud is washed off.

RECIPES FOR SUCCESS

Sacred baths are the meeting place of three energies: human, plant, and universal. To our eyes, their effects can be subtle at first, because we have not been taught to view the world or ourselves from a perspective of energy, but to see everything as physical matter. The bath is therefore like a new language of health, or a new way of seeing. Its power enters us slowly at first, as if the intelligence it contains is gently and gradually opening us and building our strength. Then, after a while, we can receive deeper healing without being overloaded or shocked. If we keep faith with the process, we will master the language of the plants and more powerfully experience the effects of the bath as time goes on.

The following recipes are offered so you can explore this yourself. The first two are Haitian in origin, the next is from Peru, and the final bath is based on the symbolic and energetic powers of homegrown American plants.

In Haiti and Peru, baths are taken outside in nature, but, for practical purposes, we suggest you use your indoor bathtub, although you should still approach bathing in a sacred manner. You can carry a ritual candle as you enter the bath if this helps you to connect with your spiritual intent; if not, simply bathe by candlelight, and state your prayers before you enter the tub.

Stand in the bath and, turning counterclockwise and then clockwise, visualize negative energies leaving you and your body filling with positive energy as you pour the water over yourself. The water, in all cases, should be as cold as you can bear.

Love Bath (Haitian)

For purifying energies and opening to love

To make this bath, you will need the following:

- A few drops of Florida Water (or perfume)
- Some sprigs of fresh basil
- A handful of rose petals

- A few cuttings of fresh aloe
- Four or five orange leaves

Steep all of the ingredients for an hour or so in hot water and, when cool, add this to a cold-water bath, along with three drops of vanilla essence and a cupful of single cream or milk. Pour the water over yourself and ask the spirits of the plants to invest you with their energy and empower you to draw in the love you need.

When you are finished with your bath, take the plant remains into nature and bury them at the base of a tree, or throw them into a stream, leaving an offering of thanks to the spirits of that place.

⸂ Power Bath (Haitian)

For the energy to deal with specific problems

This bath works directly with the power of the Lwa, who are particular spiritual energies with a specific purpose or attunement to human beings.

In Haiti, the basic bath consists of the following:

- Holy (or spring) water
- Florida Water
- A selection of the following herbs (not all are necessary): *acacia jane* (in the West, use mimosa), *amwaz* (motherwort), *balai-dou* (licorice), *balisier-rouge* (plantain), *basilica* (basil), *bwa-major* (pepper)

Depending on the nature of the problem or the spiritual intervention required, you may then add any of the following:

- Carnations and/or sandalwood. These are plants sacred to Legba, who is the gatekeeper to the other Lwa (much like St. Peter to the saints) and to the entire spiritual universe. Carnations and sandalwood will therefore connect you to the spirit world at large and bring you its fortification and strength.
- Lavender and/or rosemary. These are the plants of La Sirene (the siren), the mermaid spirit who will strengthen your emotions and

unconscious powers and give you access to the riches to be found in the deep oceans of the self.

- Mango leaves, coltsfoot, comfrey, and/or mint. Ogoun, the warrior Lwa, is called upon for power and self-confidence and is reached through these plants.
- Cinnamon, (pink) rose petals, and/or marigolds. These are for Erzulie, the spirit of love, who brings gifts of luxury, opulence, and comfort as well as harmony in romantic and other relationships.

In the African Yoruba system, which informs aspects of modern Vodou, the Lwa—here known as Orisha—are largely the same, although their herbal correspondences differ. For the Yoruba, the following herbs (known as *ewe*—pronounced "you-WAY") are sacred to each spirit, and these can also be used in your baths:

- For Obatala (the spirit known, in Haitian Vodou, as Bondye; the Creator and ultimate healer), herbs include skullcap, sage, kola nut, basil, hyssop, vervain, willow, and valerian.
- For Oshun (in Haiti, Erzulie: the spirit of sensuality, beauty, and grace), herbs include dock, burdock, cinnamon, damiana, anis, raspberry, yarrow, chamomile, lotus, and echinacea. Oshun has a particular affinity for women and babies and can be appealed to for help in childbirth, the alleviation of specifically female disorders, and for the protection and well-being of children.
- For Yemoja (in Haiti, La Siren: the "mother of waters"), herbs include kelp, cohosh, dandelion, aloe, mint, passionflower, and yams. Yemoja is concerned with sexuality, the unconscious, the emotions, and is also a protector of women.
- For Ogun (in Haiti, Ogoun): the "god of iron," herbs include eucalyptus, hawthorn, parsley, and garlic. Ogun is also the spirit of warriorship and power. He "clears the road" by making problems vanish so that energy can flow.
- For Oya (in Haiti, Brigid): the "mother of the cemetery" and the spirit of rebirth and new beginnings; herbs include mullein, comfrey, cherry bark, horehound, and chickweed.

- For Elegba (in Haiti, Legba): the messenger of the gods and gate-keeper to the Lwa, all herbs may be used since Elegba enforces the divine law of the universe, brings deeper spiritual connection, and has power over the spirit of all plants, amplifying their healing effects.

🪶 Soothing Bath (Peruvian)

For peace, good fortune, and harmony of the soul

This bath combines plants that have soothing and purifying qualities:

- *Aire sacha* (balm), which in Peru is known as the "miracle leaf" and brings inner and outer balance to body and spirit by removing mal aire (bad air; curses or negativity, often sent through envidia)
- *Comphrey* (comfrey), for protection and luck
- *Rosa sisa* (marigolds), a plant that restores the soul, and which has already been mentioned many times in this book for its soul-healing powers.

Add handfuls of each to your bath and rub them over your body as you soak.

🪶 The Language of Flowers (Traditional American)

A recipe personal to you

This is an opportunity to create your own bath according to your personal needs. It is based on the American Floral Vocabulary, or "language of flowers," which was popular in the late 1800s and recorded the symbolic meanings and attractive powers of many common American flowers. This vocabulary was once known to all young women and became a form of communication between friends, who would send flowers to each other as a "code of sentiment" and a form of blessing. The flowers would then draw into the recipient's life the qualities they represented.

Some of these qualities are listed below, and from these you can devise your own baths. The language of flowers is now largely unknown in America. By using these baths, you might also rekindle interest in this homegrown form of folk healing and well-wishing.

- Almond: Hope
- Bay: Glory
- Buttercup: Wealth
- Cowslip: Grace
- Dahlia: Elegance
- Fennel: Strength
- Heliotrope: Loyalty
- Lily (white): Purity
- Magnolia: Blessings of nature
- Olive: Peace
- Poppy: The power of dreams
- Rose (red): Love
- Star of Bethlehem: Reconciliation
- Strawflower: Unity
- Sunflower: Spirituality
- Tulip: Influence over others
- Violet: Faithfulness
- Witch hazel: Magical powers

Add to your bath the flowers representing the attribute (or attributes) you want to draw into your life and bathe in the normal way.

7

THE SCREAM OF
THE MANDRAKE

When a storm approaches thee, be as fragrant as
a sweet-smelling flower.

JEAN PAUL RICHTER

The root of the mandrake is shaped like a human being, complete with arms and legs. A powerfully magical—some say, dangerous—plant, legends tell that when it is pulled from the soil, it emits a scream of pain and rage at its violation and that of the Earth. Anyone hearing this scream is himself sure to die unless ritual precautions are immediately taken to honor the plant and its environment.

Myths and legends like these are often teaching stories and behind them lies a deeper truth. In the scream of the mandrake, there is a lesson for us about the need to "walk lightly on the earth"—to treat our world with respect, awaken to its spirit, and take only what we need. The human race and its "measuring mentality" has not been too astute at this in the past, leading to a situation of increasing climate change, which some scientists bluntly state is now irreversible. We will simply have to get used to and prepare for wilder and more chaotic weather— ferocious hurricanes, flooding, rising sea levels, blistering summers, frozen winters, failing crops, and scarcities of food and drinking water, which will increase in intensity year-on-year. We can, however, give more respect to the Earth now by cutting back on our resource plundering and pollution so that we bequeath a world with some comforts left for our children instead of a barren planet. Or they, too, may hear the mandrake scream.

THE DEATH OF A SHAMAN

Every time a shaman dies,
it is as if a library burned down.

MARK PLOTKIN

There are nearly 270,000 species of flowering plants on Earth, and less than 1 percent of them have been studied for their healing properties. Moreover, most of the research that is taking place is conducted in Western laboratories, where scientific rather than spiritual methods are, of course, employed. The intention is to isolate one or two active ingredients and patent more drugs instead of finding more cures. And as part of this process, researchers sacrifice the plant itself. It is just as if we were killing a human being to take only the teeth and hair. Any other secrets that plant might teach us die with it on the altar of Western rationalism.

Around 125,000 species—almost half the plants on Earth—are found in tropical rainforests, which cover almost eight billion acres of the world's surface One in three plant-derived drugs come from these rainforest plants yet only a fraction have been investigated for medicinal purposes.

Estimates vary, but it is well-known that several thousands of these rainforest acres are destroyed each year by Western companies or local farmers under Western sponsorship, so that cattle-grazing and mineral exploration can take place, in the interests of fast food and petroleum companies. There is no doubt that many of these disappearing plants hold the keys to lifesaving new medicines—we know this from the less than 1 percent that have been studied—and yet every year thousands more are destroyed. Once they are gone, they may never return.

But this is only half the story, because traditional ways of working with plants are also dying out as the West exports, not only its technology and needs, but its worldviews and values to these cultures. It is a frequent lament among Amazonian shamans, for example, that many fewer young people are now coming forward to learn natural medicine and to meet the spirit of the plants. They are migrating to the cities instead, or putting their faith in Western science, which sees their shamans as outdated, misguided, or a throwback to a naïve age.

These shamans, who cultivate their successors through apprenticeship, have no more students to teach, and their knowledge is dying as quickly as the forests around them. "I have always said to my children, you can be what you want, but don't forget your culture," says Guillermo Arevalo. "Come back to nature and your people." Sadly, not all of them do, despite the pleas of the medicine men themselves.*

This is a tragedy not just for Amazonian culture but for all of humanity, since many of the drugs we use in the West are derived from shamanic knowledge. For decades, pharmaceutical companies have employed anthropologists and ethnobotanists to work with these shamans so they know where to look for the plants and what they are used to cure.

In Haiti, too, the situation seems somewhat similar. On this Caribbean island there is less of a temptation for the young to adopt Western ways, since all they have really experienced of the West, in a culture born in slavery and subjected to exploitation ever since, is Western oppression. But still, what the people aspire to is often power more than spiritual communion, and this is measured in Western terms, so that money and material possessions become the new gods. This is not so surprising, since the Western use of force to take power from others is what has been taught them from birth.

---

*Encouragingly, though—perhaps even ironically—the increasing interest in traditional plant medicines by Westerners, such as those we take on our Amazonian retreats (see the About the Authors page), is now helping to revive the interest of younger Peruvians in their own traditions. Our friend the shaman Artidoro remarks, for example, that "When our children see Westerners coming here and wanting to learn about our plants and medicines, they think, 'Maybe there is something in this after all,' and they ask their elders to teach them about the plants."

As a consequence, although shamanic initiation still takes place, for many this becomes a way of earning money in an otherwise deprived country—a job more than a calling to heal—and the older and more experienced shamans talk of a decline in the spiritual power of the newcomers to their profession. "Their heart is not in it," they say.

In the heart of our own Western culture we can also chart the decline of traditional healing as science has come ever more to the fore. In the United States, many true healers and shamans live on the reservations, so their knowledge and expertise is hardly known to the wider world. And many age-old natural cures have been made illegal, along with the plants themselves. In Europe, herbalists are under increasing pressure to become educated, validated, and registered in the same way as scientists, leading to a decline in intuitive healing and knowledge about plant spirits. Throughout the world, the old ways are being denied or forgotten.

It is sad, oppressive, and potentially dangerous when cultural diversity and freedom of speech and belief are restricted in this way, but it is also self-defeating for those who oppress. Every plant is a complex mixture of interacting energies and healing processes, with at least thirty active ingredients in each one. A Western pharmaceutical drug created from any plant is lucky to contain only two or three of these active components, so we are all missing out on healing and making ourselves weak instead of well.

In his foreword to Patrick Logan's *Irish Folk Medicine*, Sean O Suilleabhain of the Department of Folklore, University College Dublin, remarks that

> The therapeutic effectivity of popular medicine cannot be judged solely from the viewpoint of modern practice. The early concepts of disease and the aims of ritual healing were quite different from those of our own time. A study of folk medicine must enquire whether it was really effective in its own environment; whether the folk healers of earlier generations were able to treat successfully the same ailments as modern man suffers from and which are now treated in a completely different way; and also

whether folk remedies were especially suitable for certain diseases or groups of diseases.[1]

Not only were they, he concludes, but he offers examples of Western cures that could not have existed at all without the discoveries of traditional healers who worked in concert with their plant spirit allies.

> African medicine men have for a long time used the bark of a certain type of willow to cure rheumatism with salicyl; the Hottentots knew of aspirin; the natives of the Amazon River basin used cocillana as an effective cough-mixture, and curare, which they applied to arrow-tips to stun their enemies, is now used as an anaesthetic; the Incas have left us cocaine; ephedrine reached the Western world from China; cascara was known to the North American Indians; from the juice of the foxglove was derived digitalin for heart ailments; and finally, here in Ireland, molds from which penicillin has been derived were traditionally used for septic wounds . . . early peoples used compresses, scarification, hot baths *(tithe alluis)*, even vaccination.[2]

Furthermore, if humankind is spared from what the media is now calling the "modern plague" of avian flu, which as of this writing had claimed sixty lives, it may not be through the marvels of modern synthetic science, but through a simple fruit—*star anise*. According to an article in the London *Independent* newspaper, this is "the only defense the world currently has against the threatened flu pandemic."[3] Star anise, the fruit of a small oriental tree, is a natural source of shikimic acid, which is the basis for the only effective antidote to avian flu. Interestingly, this flu strain began in China, which is where star anise grows, giving credence again to the shaman's claim that plants grow where they are most needed.

It seems dishonest and ungracious, at the very least, then, for modern medicine to take so much from the old ways and then belittle these traditions for their "primitive beliefs" and "lack of effective medicine."

To do so is a revelation of ignorance. As science wins the war against tradition, the old ways die out, leaving fewer folk healers and plant experts whose old knowledge our scientists can raid to develop their new medicines. One of the reasons that Patrick Logan gives for writing *Irish Folk Medicine* at all, in fact, is to document a tradition that has almost entirely been lost.

Logan, a medical doctor, makes another not insignificant point when he says that "almost all physical illnesses—over 80 percent of them—will get better *no matter what treatment is given*" (our italics). The important thing is "reassurance"—reconnecting patients with their body's own healing powers and giving them back their balance.

Patients are much more likely to receive reassurance like this in traditional healing sessions where healers give their patients hours, days, or weeks of genuine care and compassion. This is in stark contrast to the modern medical setting where, due to the demands of an increasingly unhealthy population and the time constraints on our doctors, the average consultation now takes less than ten minutes. It is surprising that anyone gets well at all these days.

And, in fact, do people get well? Some observers don't think so. Kevin Trudeau, in his book, *Natural Cures "They" Don't Want You to Know About,* contends that, for at least the last hundred years, a drug has *never* cured *anyone*.[4] Indeed, he argues, if you think about it, there is no motivation for drug companies to find actual cures, since their profits come from sick people who buy the medicines they market. Their drugs are designed to alleviate suffering so that the "customer" (not the "patient") will buy the product again. Offering a cure would be crazy; it would put the company out of business. The only thing that is absolutely required of any chief executive officer heading up a publicly listed drug company, Trudeau reminds us, is to make a profit for shareholders, and in his opinion, this actually runs counter to the idea of curing illnesses.

Of course, this attitude contrasts sharply with the approach of traditional healers, who were often paid to keep patients well and received no payment at all when their patients got ill. In China, for example, villagers paid the barefoot doctors, herbalists, and medicine men for every day they remained healthy. If a patient became ill, however, then the healer worked for free in order to cure him, only receiving payments again when the patient had fully recovered.

Furthermore, despite the grand claims of drug companies, their products may actually cause illness rather than cure it. A reporter on the London newspaper *The Independent* revealed this: "The Medicines and Healthcare Products Regulatory Agency (MHRA)—the government group responsible for regulating UK medicines, including herbs—says that between 2000 and August 2004, there were 451 reports of suspected adverse reactions involving herbal preparations, of which 152 were serious." On the face of it, this sounds like a lot. But the journalist continues: "By way of comparison, consider this from a report in the *British Medical Journal* last year: 'In England alone,* reactions to drugs that led to hospitalizations followed by death are estimated at 5,700 a year and could actually be closer to 10,000'. Herbs may not be completely safe, as critics like to point out, but they are a lot safer than drugs."[5]

There have been some spectacular drug failures, despite the promises. In 2005, for example, the FDA, after an extensive review of hundreds of studies, issued a warning that the use of antidepressants may actually lead to an *increase* in depression and suicidal thinking—the very problems these drugs are supposed to cure. The Yahoo! news story that covered this warning noted the FDA's concerns that "antidepressants may cause agitation, anxiety and hostility in a subset of patients . . . psychiatrists say there is a window period of risk just after pill use begins, before depression is really alleviated but when some patients experience more energy, perhaps enabling them to act on suicidal tendencies."[6]

By contrast, the herbal cure for depression, St. John's wort, has never harmed anyone. As the *Independent* newspaper article put it: "[St. John's wort] is not only more effective in the treatment of moderate to

---

*That is, not including Scotland, Wales, or Ireland.

severe depression than the SSRI Seroxat, according to the *British Medical Journal*, but it also has fewer side effects."[7]

The latest "miracle drug" to cure arthritis, Vioxx, also ran into problems when it was discovered that one of its side effects was to double the risk of heart attacks. Aspirin-like drugs (NSAIDs) used for the same purpose have been estimated to cause 2,600 UK deaths a year as a result of intestinal bleeding. Once again, the herbal alternative, garlic and devil's claw, has never harmed anyone. The *Independent* article does report: "A study expressed concern about herbal remedies that could interact with treatments like NSAIDs . . . leading to increased gastrointestinal bleeding." But, as the journalist points out, "the herbs don't cause the bleeding, it's adding the aspirin."[8]

In the face of all this evidence, the only sane conclusion is that it is time to do things differently: It is time for a return to traditional, compassionate healing methods, to concern for patients instead of for profits. And it is time for a new generation of plant spirit healers—you, the readers of this book—to step up to the plate to arrest this decline in the well-being of the world.

To do so, it will be necessary to free your minds from the conditioning of scientific rationalism, so you can explore, dream, meet, and work with your plant spirit allies, the energies of nature that are calling you. If enough of you answer this calling to rediscover the magic of plant spirit healing, together we can preserve the traditions and use them for the good of all, altering the course of this increasingly materialist and dis-spirited world.

> *It is time for us to kiss the earth again,*
> *It is time to let the leaves rain from the skies,*
> *Let the rich life run to the roots again.*
> ROBINSON JEFFERS

In the remainder of this final chapter we offer some suggestions as to where your explorations might take you, as well as a few more uses for nature's great healers.

HEALING OUR ANIMAL BROTHERS

Some people are surprised that plants work as effectively with other animals as they do with humans, but there is really no mystery to this. As the evidence shows repeatedly, plant spirits have an affinity for all life. They are as aware of the presence of a spider in a room as they are of a human being, and they can communicate with both just as effectively, in a nonverbal language of their own.

Anthropologists tell us, in fact, that if plants were the first medicines, then animals were the first doctors, knowing naturally which herbal species to eat to cure their illnesses and discomforts. It was through the observation of other animals, these anthropologists suggest, that human beings learned to discriminate between the plants for different ailments and began to develop their own herbal knowledge. Modern studies of gorillas in the wild, for example, demonstrate a sophisticated understanding of plant remedies; and when human beings have tried the same herbs for the same conditions, they have worked just as effectively on us too.

Because of this affinity between animals and plants and the natural talent of animals to doctor themselves, some shamans trust them faithfully to prescribe medicines for their patients.

One Andean method, for example, is for the shaman to journey to his plant allies and, based on what they tell him, to select a number of possible herbal cures. He arranges these in several small piles in front of the patient. The final prescription is left to a guinea pig, however, which the shaman takes in his hand and rubs gently over his patient's body so the guinea pig absorbs an awareness of the ailment and its probable cure. The shaman then places it on the ground, and whichever pile of herbs it runs to and begins to nibble is the medicine the shaman will use.

In the jungles, too, the spiritual connection between plants and animals is well-known. Shamans, in their ayahuasca visions, are aware of this shared energy, so that the jungle itself becomes a living, breathing entity that moves in the way of an animal. Vines become snakes, bushes become puma, and the rain on the leaves holds dolphin energy.

In *Ayahuasca Visions*, Luis Eduardo Luna records the shamanic view that there are plants whose ancestors were animals. Of the vine *bejuco de las calenturas* (vine for fevers), for example, he says, "Its 'mother' is a boa . . . Mixed with ayahuasca it is most powerful in curing extreme cases, patients with convulsions and typhus [convulsions, of course, being reminiscent of the writhing of a snake]. . . . There is another [plant], *shillinto blanco* (white shillinto), whose mother is the rainbow."[9]

Since shamans make no distinction of status between plant, human, and animal, it is also not uncommon for them to share the spirit of their visionary allies with animals. Thus, "Lamista Indians give ayahuasca to their dogs before hunting. . . . Among the Jivaro, dogs are given the hallucinogenic Datura (*Brugmansia*) to help them obtain supernatural power."[10]

Once upon a time in the West, we also knew the power of plants for healing our animal brethren. *Irish Folk Medicine* lists numerous natural cures for horses, cattle, sheep, pigs, and dogs, all of which were treated with respect as companions and equals as much as chattels.

In the Bunratty Folk Museum, for example—a "living museum" between Shannon and Limerick in Ireland—there is a fine example of a "byre dwelling" from County Mayo, which was a traditional cottage occupied by both people and their milking cows. The bedrooms are on one side of the single-story building, the sleeping area for the cows on the other, and the living space was shared by both. And in Dingle, there are numerous examples of houses where the hearth fire was shared by the family and its pig, which, as a sacred animal in Celtic lore, was rarely consigned to a sty.

Nowadays, these healing traditions and the respect for animals that went with them has largely given way to commercial and factory farming; thus, "veterinary folk medicine is almost forgotten," says Logan. "In order to learn about the treatments used it is necessary to ask elderly men who could have used the treatments; few young men, unless they have seen their fathers use them, know anything about them."[11]

Among the cures that a fortunate young person might discover if he or she were to ask an elder, are horse parsley to calm an unsettled horse, furze bush tops and linseed oil to expel worms, and a dressing of *an pocan* ("buck mushroom" or "puff ball") spores to stem blood flow from a cut (this also works for humans).

To relieve cows of "the bloat" (a painful accumulation of gas in the stomach), Logan records a cure of ginger, wood charcoal, *nux vomica*, gentian, and baking soda, which is given as a thin gruel every twelve hours.

Sheep were treated for the common disease of liver fluke with iris flowers, wallflowers, and ragwort, all of which are yellow, another example of the doctrine of signatures. Liver problems will produce a yellow hue in the sufferer and can therefore be cured by administering plants of the same color.

Other natural cures include the following:

- Hops, sweet cicely, and lady's mantle as a tonic for horses, sheep, and goats
- Nettles as a mineral and protein-rich forage for horses, and for washing the coat to give shine
- The acorns of the oak, ground and mixed with wheat flour to stop bleeding (this can be used internally and externally)
- Black poplar buds in milk and honey to soothe ulcers in horses
- Wild strawberry leaves to strengthen animals during pregnancy
- Willow, to cool the blood and calm the nerves in goats

All of these cures came from the quiet observation of animals in nature and the gentle attunement of keepers and healers with the natural world and the spirit of the plants. Such methods require nothing more than that we slow down to the pace of the Earth and break away from our modern obsession with frenetic activity. Then we find that nature itself becomes our university, and we begin to see the spirit in all things.

With animals, of course, it is most unlikely that any "placebo effect" is in play. If the cure works, it works. Our plant allies are simply happy to provide healing to those who ask. And if such cures work with animals, they will work for us too.

A FRAMEWORK FOR NATURAL HEALING

From your reading and your own explorations with the spirit of the plants, perhaps you are inspired to work more closely with natural healing methods and, one day, to extend your practice to working with patients of your own. We would like to offer a few words on this, as there is an effective and proven framework for healing used by many plant spirit shamans that may be of value in your own consultations.

The first thing to know, however, is that most Western patients these days have little idea about shamanic healing and may even be wary of talk about plant "spirits" and "otherworld allies." This is part of the sadness of our times: many of us have become so out of touch with nature that we are now even scared of it.

The fact that the patient is anxious or afraid and still coming to see you despite his or her trepidations is indicative, however, of the deep need for healing that people are feeling in our unbalanced age.

On the way to you, because of his anxieties, your patient may feel strange, and more aware of himself and his emotions. This is exactly what you want. When you perform a healing, you are not just handing out a drug, you are asking the patient to meet you by shifting his consciousness into a slightly altered state so he can receive healing; and he is already beginning the process during the journey to you.* At the very least, in this frame of mind, he will be able to talk more easily about feelings and fears because he will be in a different mental space from "normal." Your intention as a healer is to build on this so he can relax further into his true self and emotions.

Prepare your healing room with this in mind, and try to create an "otherwordly" feel so you are supporting the subtle shift in awareness. This may mean incense and candles (more relaxing and healthier than electric light anyway), ritual items, power objects and, of course, healthy growing plants—anything that creates a nonordinary impression without veering into flakiness. Medical doctors do this all the time—their "props" are white coats and stethoscopes, books on human anatomy,

---

Altered, in this sense, means free of the everyday trance we are all socialized into and which has become our habit.

their medical degrees and qualifications on the wall; and traditional healers also display their medicine tools, probably for the same reasons.

There are also some things you need to prepare in advance that
aren't props, but essential tools. Make sure you have these items:

- A blanket and pillows for your patient to lie on
- Smudge mix or incense, and a holder for this
- A lighter or matches for the smudge
- Your chacapa (or rattle)
- Your drum, drumming tape or CD, or whatever you use to shift
  into otherworld consciousness

You also need to feel full of power, especially if you're likely to be
dealing with spirit intrusions (see chapter 4). There are a number of
ways to call in power, all of which traditional healers use:

- Journey to your plant and nature allies and ask for their help with
  the healing.
- Drum, dance, or sing your plant song to establish a spirit connection.
- Sit in meditation before your altar and consult with your seguro,
  your plant ally, which is a gateway to nature (see chapter 1).

Whichever of these methods work best for you, these are the ones to use.

Welcome your patient when he or she arrives and begin by smudging
him, yourself, the room, and the tools you will use. As you cleanse each
one, ask for its help in the healing. Everything is alive and has the right
to be asked and invited in.

Then you may talk for a few minutes to set your patient at ease. This
is what the Andean curanderos call a *platicas*—a heart-to-heart, soul-to-
soul consultation. As you talk, you should also watch and listen closely
to your patient. This may give you further clues about the problem he
needs help with.

Obvious examples include the following:

He says: "I really love my mother . . ."
Listen for the but, even if he never says it. Is it there somewhere,

spoken silently? Might this suggest a problem with power or even soul loss?

She says: "I've never been happier than in my current job."

But notice how she wrings her hands as she says it. What might that mean?

Where do his eyes go?

Up and to the right: he's likely to be remembering something. Up and to the left: he may be imagining something, or is not sure about it, but it might have happened, sometime, somewhere, once . . .

How is she breathing?

High up in her chest may suggest fear as she talks of a particular event (fear probably means an issue to do with power loss), low in her stomach may suggest grief (and this may imply the need for a releasing ritual). You will form your own relationship with the breath as you continue your work and come to understand what these clues mean to you.

These and other observations will help you to help him or her by getting an intuitive sense of the problem, which a patient may not be able to vocalize.

While you are conducting this platicas, also use your powers of *gazing* to *see* your patient. Allow your eyes to go slightly out of focus and use your peripheral vision to look just past him. As you do so, what do you see in his energy body? What images enter your mind about what might be going on for him? Trust your first impressions. If you start telling yourself "That can't be right," your rational mind will get in the way, and that's when you can lose your connection to power again.

After you have chatted for a while, take the patient into your healing space, let him lie down, and relax him. Make sure your hands are clean, as you are about to touch his soul, literally. Agua florida or oils of frankincense, jasmine, geranium, sandalwood, or myrrh are good purifiers and can be simply rubbed on your hands. Then begin your healing work, which may include spirit extraction or power and soul retrieval. In all cases, your plant spirits are your strongest allies and will delight in helping you if you remain open and listen for their voices.

Shamans are "walkers between worlds," and it is usual for them to carry out a healing in ordinary as well as nonordinary reality, so that a beneficial change is made to the patient's energy field and plant medicine is also then administered in the physical world.

In our experience, this combination of physical and nonphysical healing is always more effective than just transferring or transforming energy. For example, a few months ago, one of Ross's patients came for a consultation to help alleviate her problems with ME (myalgic encephalomyelitis; i.e., chronic fatigue syndrome), and he did energy work to relieve her discomfort and tiredness.

During the journey, however, the spirits also recommended that she take a tea of horehound, comfrey, or lungwort, plants that would not normally be prescribed for ME. Nevertheless, this information was reported to the patient, who said she would take these herbs.

She returned for a second consultation a few weeks later, and this time she was suffering from a severe cough, which, with her other condition, was proving debilitating. Asked if she'd taken the tea that the spirits had advised, she replied, "No. I didn't have time to find the herbs." Which was a pity—since horehound, comfrey, and lungwort are some of the finest herbs for chesty coughs. If she had followed the recommendations of spirit, she might never have caught an infection at all. Proof again that it is a foolish healer (and patient) who doesn't act on the suggestions of the spirits; proof also of the value of combining plant medicines with energy healing.

When the healing is complete, your spirits should, as a matter of course, be thanked for their help and released. Then you can seal the energy body of the patient by rattling around it with a rattle or chacapa. If your plant allies have offered any guidance or counsel, this might be the best time to relay it, so the patient can make any choices or changes necessary or write down any herbs prescribed, if you will not be preparing these for him yourself.

EXTERNAL MEDICINES

For those patients who, for whatever reason, prefer not to use herbs internally, or where the nature of the illness suggests external application, there are other ways for you as the healer to pass on these plant medicines.

You can easily prepare a cream, for example, by adding an ounce of warmed olive oil to an ounce of beeswax and half an ounce of lanolin, along with half a teaspoon of borax. To this, add the liquor of half an ounce of herbs (chosen according to the nature of the problem or the spirit advice you have received), after the herbs have been steeped in a half-pint of boiling water for about thirty minutes. You can add a few drops of perfume or Florida Water if you desire; and then, of course, the mixture should be blown with tobacco smoke so your intention is contained within it. When the mix starts to thicken, pour it into a pot and refrigerate. It will last some months when stored this way, and the borax will help to preserve it.

An even simpler recipe for a cream is to melt petroleum jelly in a saucepan and stir in herbs that have been thoroughly crushed with a mortar and pestle in order to release their essences. After a while of gentle warming through, take the saucepan off the heat and let the liquid cool. Made in this way and kept in a refrigerator, it should last for up to six months.

Eyebaths can be made by adding a teaspoon of suitable and soothing herbs (such as fennel, rose, aloe, fenugreek, periwinkle, elderflower, or slippery elm) to a mug of boiled water and allowing them to steep for about five minutes. When cool, apply the infused water to the closed eye with a cotton wool pad, and invite the patient to relax for five or ten minutes as the plant energies soothe her eyes.

Similarly, you can make natural soaps and detergents by adding about three ounces of herbs to a pint of hot water and steeping them for ten minutes. Then sieve the solution and decant into a jar. Good detergent herbs include feverfew, marigold, chickweed, daisy, and goldenseal. The best herb for soaps is soapwort, which, as the name implies, was used for cleansing before the invention of chemical substitutes. All of these are natural alternatives to synthetic products and are especially useful for patients with allergies or eczema.

You can use plants as aromatic oils in a burner as well, of course, and, in this way, they need never make contact with the skin at all. One rich Hoodoo recipe for cleansing the aura of negativity is prescribed in just this way. It consists of ten drops each of patchouli, lotus, carnation, and gardenia, five drops of frankincense, myrrh, eucalyptus, and lavender, and two drops of cinnamon, added to plain water, which is heated with a candle flame.

If desired, this Hoodoo blend can also be used in a bath by mixing it with four ounces of grapeseed oil and adding it to the water. Invite the patient to soak in the bath for fifteen minutes, visualizing his limitations and negativity leaving the body. He then stands, leaves the tub, and watches as the water drains away, carrying the unhelpful energy with it, so he feels renewed and refreshed.

And, of course, herbs can be used in their purest form, without preparation at all. Lavender, hung above a bed, will aid restful sleep; vervain and valerian, hung in the house, will protect against unhelpful energies; and angelica stalks hung in a car will prevent travel sickness. In fact, more elaborate preparations are rarely necessary, as the spirit of the plants is all that you really ever need.

Prescribing Plant Cures

The following exercise suggests one approach you might take to consult with your spirit allies in order to prescribe herbal remedies for your patients, whether animal or human. This method has little to do with modern medical herbalism, in which, through normal face-to-face consultation and diagnosis, the practitioner prescribes a particular herb to treat a particular symptom, based on the chemical properties of that plant. Ours is a more traditional approach, which works by forming a holistic connection between the herbalist, the patient, and the spirit world, and which gives power back to the patient by involving him or her in her own healing.

Close your eyes and see before you the world tree, the *axis mundi* at the center of the natural universe, and, next to this, the entrance to a cave that leads into the Earth. As you stand before the tree, call your intention to mind: *I am here to meet with an otherworld healer and take advice on the plant medicine most needed by my patient.* When you are clear on this, enter the cave, and then follow it down into the Earth.

As you proceed, see that you have your patient with you and that you are escorting her on a journey toward healing. Take note of how she behaves in the cave. Does she move easily and fluidly toward healing, or withdraw from it and need to be led? Is she excited to be here and hopeful of a cure, or fearful and hesitant, relying on you to guide her? This spiritual or intuitive information may be useful to you and your patient in identifying the patterns underlying her illness and her obstacles to health. These observations can form the basis for a subsequent platicas, so remember what you notice here and allow all your senses to inform you.

Eventually you will come to the end of the tunnel and step out into the otherworld. The terrain will be unique to you, and it may also be influenced by the presence of your patient, who will have her own otherworld "dream." But certain features are the same for everyone, and it may be that you find yourself standing on a plane, with a sea or lake to your right and a jungle or forest to your left. Move toward the forest, and you will come to a small wooden bridge across a shallow stream. Just beyond the bridge, toward the right, is a lone jungle hut made of branches and thatched with reeds. This is the hut of a healer.

Look around you as you enter this simple hut with your patient. The floor is of earth and straw, the walls plastered with clay and mud; shelves contain pots of herbs and leaves, and the only items of furniture are a small wooden table and a chair on which the healer sits. He rises to greet you and you notice that he is dark skinned, dressed simply, a short man, elderly you suspect (though his age is hard to tell), and he appears friendly, wise, and powerful—a perfect healing blend.

On this initial journey, you have two purposes: to find healing for your patient, of course, but also to introduce yourself to this spirit-healer who will become your ally. So talk to him, tell him who you are, what you are here for, ask for his name (some know him as Kinti, but to you it may be different), and request that he help you and your patient, which he will willingly do.

After this journey, it will serve you to return, this time without a patient, and meet this healer again. He has a vast knowledge to impart and you will find that, in time, he will also take over much of the healing and teach you through example and explanation. Being in his company then becomes as rich and valuable as an intensive course of one-to-one instruction with an expert herbalist and physician.

For now, though, the need is your patient's, so introduce her to this healer and explain her symptoms of ill health. Then step aside so you can watch and learn.

Kinti's healing methods are based entirely on plant medicines and magic, and he will use many of the techniques now familiar to you: cha-capas to change energy, perfume sprays to "flourish the soul," incense to cleanse, and floral baths to refresh and revivify. But it is the depth of his skill in using these tools and his ability to see the illness of a patient in her energy field that is most impressive. His other skill is in advising on the herbs the patient should take in ordinary reality. He will consult with you on this, so listen carefully to what he says.

When the healing is complete, thank your ally, and escort your patient back to the everyday world, retracing your steps through the cave and noticing, once again, her reactions as you move along the tunnel. Most often she will be much calmer, and any obstacles you encountered on your first passage will now be removed, suggesting a more harmonious and balanced relationship between your patient and the Earth, which is the outcome of all successful healings.

When you are back in ordinary reality, advise your client of the herbs recommended to her and also the method by which she should take them (as a tea, a bath, a cream, or in aromatic oils, etc.) and for how long. All of this information will have been given to you by Kinti.

It is amazing, actually, how accurate this healer is in his diagnoses and prescriptions, and how effective his treatments are. Since meeting him some years ago, Kinti has assisted in hundreds of our healings and has often prescribed herbs that we, as healers, would never have thought of for a particular patient, and even some that we had never used before.

An example of this was one of Ross's patients, who presented with a rare bladder condition. Kinti prescribed "stone root," an herb that is hardly used in the UK and which Ross had no prior experience with. Looking it up in an herbal encyclopedia, however, it turned out that stone root (named for its effectiveness in curing kidney and gall stones) is also a great bladder healer and recommended for many conditions of the urinary organs.*

Experiences like this, where information is received that the human healer did not and could not otherwise know, are proof of spirit intervention in the healing and examples, too, of why a relationship with this otherworld healer is so valuable.

RE-ENCHANTING THE WORLD

And so we come to the end of this book—and to the beginning, we hope, of your own adventures in healing alongside the plant spirit allies you have made.

People (none more so than scientists and clinicians) tend to over-complicate plants—their attributes and effects and the methods of working with them. There are plenty of scary stories, too, about the harm that herbal medicines may do to the unwary and "uninformed." This has left many people today fearful of nature, and of course, has ensured that its healing knowledge remains in the hands of a commercial elite. In reality, there is nothing to be scared of. People are more at risk from the "miracle drugs" of medical science than they ever will be from plants (provided, of course, that we do not "genetically modify" them or pollute the environments in which they grow, and that we, as healers, are sensible in our use of them).

Actually, plants are simple things and they have our best interests at heart, enjoying nothing more than working with their human allies to bring healing to those who request it. As you explore the great medicine cabinet of nature, you will discover this too, and so liberate yourself

---

*It is, of course, advisable that you double-check all herbal prescriptions in an encyclopedia and especially that you discuss any contraindications with your patient, who should also take advice from a doctor before any new treatment is started.

from this mindset of fear that permeates the modern world.

Furthermore, by simply working with plants in these ways—touching, tasting, and smelling them—you will have begun an even more profound process of healing: one of re-enchanting the world by allowing your spirit and that of your patients to meet the greater spirit that is behind all things. And, through this, we may all find a more honorable and holistic way of being on the Earth.

> *If we apply our minds directly and competently to the needs of the Earth, then we will also have begun to make fundamental and necessary changes in our minds. We will begin to understand and to mistrust and to change our wasteful economy, which markets not just the produce of the Earth, but also the Earth's ability to produce.*
>
> WENDELL BERRY

> *I have come to terms with the future.*
> *From this day onward I will walk easy on the earth. Plant trees.*
> *Kill no living things. Live in harmony with all creatures.*
> *I will restore the earth where I am. Use no more of its resources than I need.*
> *And listen, listen to what it is telling me.*
>
> M. J. SLIM

> *Until he extends the circle of his compassion to all living things, man will not himself find peace.*
>
> ALBERT SCHWEITZER

APPENDIX 1

A CARIBBEAN HERBAL

Here, we offer selections of frequently used herbs in Haiti, along with their direct Western counterparts or, where necessary, Western analogues for plants that are more easily found. We also list the medicinal uses for these plants as well as their magical properties, so you can continue to explore them on your own.

In this Caribbean herbal, the first herb listed in column five (in capitals) is always the direct Western counterpart (where there is one) of the Caribbean plant (in column one). Those that are not direct equivalents, but that have similar effects, are listed below it in normal script.

Medical complaints that these herbs address are listed in column three. Analogues and alternatives used in such treatments are in column five. Magical uses are listed in column four, and herbs for specific magical uses appear in column six.

Since the same plants often work for many different conditions, to avoid repetition, not all medical or magical uses are listed for every herb in every case. If you don't see your condition or need against a particular herb, try scanning the other columns and you will almost certainly find it there.

A CARIBBEAN HERBAL

KREYOL NAME	SCIENTIFIC NAME	MEDICINAL USES	MAGICAL ATTRIBUTES	ANALOGUES & ALTERNATIVES (MEDICINAL)	ANALOGUES & ALTERNATIVES (MAGICAL)
Abiaba	*Chrysolphyllum caimito*	Cancer Eye weakness, eye strain, and associated problems	Used in initiations "Opens the head"	STAR APPLE Burdock, cleavers, Echinacea, poke root (cancer) Gingko (eye problems)	Anise, bay, copal, hyssop, rosemary, holy thistle, valerian (purification)
Amwaz	*Leonurus cariaca*	Nervous disorders Heart problems	Balances the energy system Luck in love	MOTHERWORT ("Drink motherwort tea and live to be a source of continuous astonishment and frustration to waiting heirs"—traditional saying, Europe) Betony, oats, St. John's wort (nerve tonics) Lime flowers, buckwheat, hawthorn, mistletoe (heart disease)	Angelica, lemon balm, barley, gardenia, hops, rowan, sorrel, wintergreen, comfrey, marigold, fenugreek (general health and balance) Apple, balm of gilead, bedstraw, cherry, clove, black cohosh, crocus, lovage, love seed (love)
Baselle	*Ocimum basilicum*	Nervous problems Poor milk flow in nursing mothers	Overcomes fear Purifies the energy system Calls the Lwa (spirits)	SWEET BASIL Betony, oats, St. John's wort (nerves) Agnus castus, borage, fennel, holy thistle, nettles raspberry leaf (milk flow)	Gardenia, lavender, meadowsweet, vervain (peace) Carnation, ebony, ginger rowan (spiritual power)

KREYOL NAME	SCIENTIFIC NAME	MEDICINAL USES	MAGICAL ATTRIBUTES	ANALOGUES & ALTERNATIVES (MEDICINAL)	ANALOGUES & ALTERNATIVES (MAGICAL)
Bonbonyen	*Aloysia triphylla*	Asthma Colds, Fever, Gas, Stomach upset Stomach cramps Diarrhea	Brings good luck	LEMON VERBENA Elderflower, peppermint, feverfew (colds) Agrimony, shepherd's purse, sage (stomach upsets and diarrhea)	Allspice, daffodil, fern, hazel, heather, moss, rose, violet (luck)
Bwa let	*Rauwolfia seprentina*	Anxiety High blood pressure, Fever, Insanity and melancholia Insomnia (used with infants to induce sleep) Requires careful use and may be illegal in some countries	Spiritual certainty Soul healing (A note on the serpent: Damballah, the snake-god is regarded in Haiti as the progenitor of life and a great healer. He is married to Ayida Wedo— the rainbow serpent. Cf. the two intertwined snakes of the caduceus.)	SERPENTINE WOOD/INDIAN SNAKEROOT Chamomile, skullcap, oats, damiana, valerian (anxiety states) Passionflower, valerian, lime flowers (insomnia)	Coltsfoot, crocus, damiana (spiritual vision) Angelica, lemon balm, gardenia, hops, sorrel (healing)

A CARIBBEAN HERBAL (continued)

KREYOL NAME	SCIENTIFIC NAME	MEDICINAL USES	MAGICAL ATTRIBUTES	ANALOGUES & ALTERNATIVES (MEDICINAL)	ANALOGUES & ALTERNATIVES (MAGICAL)
Bwa major	*Piper aduncum*	Colds Gum disease (massage into gums) Impotency (eat 6 peppers with 4 almonds once daily in milk) Nervous upsets Amnesia (grind the pepper in honey and take once a day) Muscular pains (use in massage oils)	Spirit extraction (heats up the body to draw out negative energies)	WILD PEPPER Comfrey, sage, walnut (gums) Rosemary, ginko (memory loss)	Angelica, basil, clover, elder, fern, garlic, mistletoe, nettle, pepper, rosemary, rue, sandalwood (used in exorcisms)
Fey let pase	*Sansevieria trifasciata*	Inflamed skin conditions and soreness—eczema, rashes, shingles, etc.	"Good luck plant"	SNAKE PLANT Comfrey + Irish moss + almond oil ("skin food")	Aloe, bluebell, holly, oak, strawberry, rose (luck)

KREYOL NAME	SCIENTIFIC NAME	MEDICINAL USES	MAGICAL ATTRIBUTES	ANALOGUES & ALTERNATIVES (MEDICINAL)	ANALOGUES & ALTERNATIVES (MAGICAL)
Gingembre	*Zingiber officinale*	Nausea; Irritable Bowel Syndrome (IBS); Colds; Loss of appetite; Simple stomach problems; Poor sex drive; Stomach parasite	Spirit extraction (draws out negativity); Balances energies	GINGER; Agrimony, hops, meadowsweet (IBS); Gotu kola, ginseng, saw palmetto (sexual debility)	Angelica, basil, garlic, nettle, pepper, rosemary, rue (spirit extraction); Angelica, balm, barley, rowan, sorrel (for balance)
Kandelon	*Mimosa pudica*	Bacterial infections; Skin inflammation; Sickness, Insomnia; Irritability, Dysentery; Blood infections	Calms the spirit	MIMOSA; Echinacea, Nasturtium (antibiotic)	Myrtle, olive, pennyroyal (for peace)
Kanel	*Cinnamomum zeylanicum*	Digestive problems; Appetite loss; Colds and flu; Irritable Bowel Syndrome (IBS); Chest complaints; Lice (rub on the skin and hair)	Breaks hexes; Attracts love; Attracts power (over others—charisma, etc.)	CINNAMON; Fenugreek, gentian, milk thistle (loss of appetite); Tea tree oil or rosemary oil + peanut oil rubbed into hair (lice)	Chilli pepper, datura, holy thistle, vetivert (to break hexes); Avocado, balm, basil, betony (for love)

A CARIBBEAN HERBAL (continued)

KREYOL NAME	SCIENTIFIC NAME	MEDICINAL USES	MAGICAL ATTRIBUTES	ANALOGUES & ALTERNATIVES (MEDICINAL)	ANALOGUES & ALTERNATIVES (MAGICAL)
Lalwa	Aloe barbadensis	Sunburn and other burns Acne Skin rashes Ulcers, Eczema Shingles, Allergic reactions, Oral infections (as a mouthwash) Stomach ulcers Animal bites (as a poultice)	Good luck in love and money Protection (may be hung over doors to prevent the entrance of evil)	ALOE VERA Oats, nettles, St. John's wort (for shingles) Skullcap—also known as mad dog (for animal bites)	Almond, basil, bergamot, blue flag, dill, goldenrod, jasmine, periwinkle, sesame (for money luck) Agrimony, ash, betony, briony, carnation, cedar, foxglove, hazel, parsley, valerian, willow, wormwood (for protection)
Lay	Allium sativum	Thrombosis Arteriosclerosis High cholesterol Bronchitis Chest complaints Throat and sinus problems Hay fever Ear infections Intestinal worms	Cleansing the blood (of intrusions, etc.) Protecting the energy system	GARLIC Hawthorn, ginko, orange leaves (for arteriosclerosis) Hawthorn berries, lime flowers (to lower cholesterol) Borage, orange leaves, mint, nettles (for thrombosis)	Nettles (for blood purification) Bamboo, blackberry, cactus, dogwood, eucalyptus, frankincense, gorse, honeysuckle, lime, marigold, woodruff (for protection)

KREYOL NAME	SCIENTIFIC NAME	MEDICINAL USES	MAGICAL ATTRIBUTES	ANALOGUES & ALTERNATIVES (MEDICINAL)	ANALOGUES & ALTERNATIVES (MAGICAL)
Melon dlo	*Citrullus lanatus*	High blood pressure; Edema of the ankles	Empathy, intuition, accessing the unconscious, balancing the emotions.	WATERMELON	Cinquefoil, jasmine, marigold, mimosa, mugwort, rose (for lucid and prophetic dreams)
		Liver problems and jaundice		Dandelion, yarrow (for edema)	
		Heart disease		Boldo, plantain (for jaundice)	Bay, borage, honeysuckle, peppermint, thyme, yarrow (for psychic development)
		Kidney problems	Watermelon is offered to La Siren, loa of the seas and the deep unconscious, the emotions, etc.	Cayenne, prickly ash bark or horseradish root in honey (for heat exhaustion)	
		Heat stroke and heat exhaustion			Caraway, lily of the valley, mustard, walnut, lavender, loosestrife, passionflower, skullcap (for mental and emotional balance)
		Constipation and indigestion		Watermelon (contains more of the antioxidant lycopene than any other fresh fruit or vegetable, lowering the risk of prostate and uterine cancers)	
		Headache			
		Nausea			

A CARIBBEAN HERBAL (continued)

KREYOL NAME	SCIENTIFIC NAME	MEDICINAL USES	MAGICAL ATTRIBUTES	ANALOGUES & ALTERNATIVES (MEDICINAL)	ANALOGUES & ALTERNATIVES (MAGICAL)
Muscat	*Myristica fragrans*	Diarrhea Dysentery Colic Nausea Gallstones Headache Muscle tension in the back of the neck	Cleanses the spiritual constituents of the blood Ensures fidelity in lovers	NUTMEG Chamomile, fennel, lovage (for colic) Boldo, parsley, strawberry leaves (for gallstones)	Nettles (for blood purification) Chickweed, clover, magnolia, spikenard (for fidelity)
Planten	*Plantago major*	Blood disorders Skin rashes Diabetes Kidney and bladder problems Fevers Irritable Bowel Syndrome (IBS) Dysentery Scalds and burns Wounds	Attracts helpful spirits and protects against malevolent ones. Worn on the clothing, will also protect against animal bites	PLANTAIN Nettles, alfalfa, dandelion, peppermint (for diabetes) Echinacea, burdock (for blood purification)	For protection, see lay (garlic)
Safran	*Curcuma domestica*	High cholesterol Liver and gall-bladder diseases Rheumatoid arthritis	Purifies a space because spirits are drawn to its heat. Can therefore also be used to lead and control spirits	TURMERIC Agrimony, heather flowers (rheumatics)	Anise, bay, bloodroot, broom, chamomile, fennel, horseradish, iris, vervain, yucca (to purify)

KREYOL NAME	SCIENTIFIC NAME	MEDICINAL USES	MAGICAL ATTRIBUTES	ANALOGUES & ALTERNATIVES (MEDICINAL)	ANALOGUES & ALTERNATIVES (MAGICAL)
Shoublak	*Hibiscus rosa-sinesis*	Fevers Headache Skin problems Boils (as a poultice)	Love charms	HIBISCUS Chickweed, comfrey, marshmallow (for boils)	To draw love, see amwaz (motherwort) For divinations concerning love: mullein, pansy, rose, willow To remove love spells: lily, lotus
Sitwon (1)	*Tilia platyphyllos*	High blood pressure Hardening of the arteries Hysteria Insomnia Epilepsy Nervous tension Colds and fevers	Cuts through magic	LIME (FLOWERS) Passionflower, rosemary, valerian (for hysteria) Mistletoe, hyssop (for epilepsy)	Pepper, huckleberry, thistle (to remove or undo magic)
Sitwon (2)	*Citrus limonum*	Coughs and colds Diphtheria Malaria Rheumatism	Restores balance (especially after magical attack and spirit removal)	LEMON Peruvian bark, nettle, yarrow (for malaria)	Bay, carnation, masterwort, pennyroyal, St. John's wort, thistle (restores spiritual strength)

A CARIBBEAN HERBAL (continued)

KREYOL NAME	SCIENTIFIC NAME	MEDICINAL USES	MAGICAL ATTRIBUTES	ANALOGUES & ALTERNATIVES (MEDICINAL)	ANALOGUES & ALTERNATIVES (MAGICAL)
Sitwonnel	*Cymbopogon citratus*	Bacterial and fungal infections Nervous conditions Muscle pain and stiffness Water retention	Increases lust; makes others lust after you	LEMONGRASS Black cohosh, mullein (for muscle pain)	Avocado, carrot, cattail, damiana, endive, licorice, patchouli, vanilla, yohimbe (for passion)
Tabak	*Nicotiana tabacum*	Eye diseases (bathe in a light tobacco wash); Strokes and neuralgia (wash affected areas in water and tobacco) Body toxins (use in a bath); Earache (blow the smoke into the ear); Head lice (strong infusion in hot water poured over head); Worms (in animals—feed to them); Insecticide (also sprinkle around plants to prevent slugs)	Spirit offerings Purification Encourages spirit presence and communion	TOBACCO Plantain, gingko (for the eyes) Buckwheat, gotu kola, mistletoe (for strokes) Feverfew (for the ears)—hold the ear over a mug of the hot tea	Yew (for raising the dead) Mugwort, poplar (for astral projection)

KREYOL NAME	SCIENTIFIC NAME	MEDICINAL USES	MAGICAL ATTRIBUTES	ANALOGUES & ALTERNATIVES (MEDICINAL)	ANALOGUES & ALTERNATIVES (MAGICAL)
Ten	*Thymus vulgaris*	Respiratory illness Irritable Bowel Syndrome (IBS) Bed-wetting in children Worms Mouth ulcers Sore throat Wounds and cuts (as a compress)	Courage and energy	THYME Bearberry, cranesbill (for incontinence/ bed-wetting)	Borage, black cohosh, mullein, yarrow (for courage)
Vanille	*Vanilla planifolia*	Respiratory pain Congestion Poor circulation Minor heart problems Hysteria Rheumatism Insomnia Stress Anxiety Depression Anemia Snakebites (as a poultice)	Love magic	VANILLA Cayenne, ginger (for circulatory problems)	Aster, avens, cardamom, coriander, hyacinth (to make yourself irresistible to others)

A CARIBBEAN HERBAL (continued)

KREYOL NAME	SCIENTIFIC NAME	MEDICINAL USES	MAGICAL ATTRIBUTES	ANALOGUES & ALTERNATIVES (MEDICINAL)	ANALOGUES & ALTERNATIVES (MAGICAL)
Veven	Verbena officianalis	Postviral fatigue Chronic Fatigue Syndrome (CFS) ME; Epilepsy Jaundice Depression Paranoia Agoraphobia	Peace, calm, and prosperity	VERVAIN Ginseng, garlic, Echinacea (for CFS) Kelp, valerian (for paranoia) Sarsaparilla, gotu kola, lemon balm (for agoraphobia)	Dulse, morning glory, violet (for peace) Alfalfa, ash, benzoin, oak (for prosperity)
Vilnere	Salvia officianalis	Mouth and throat problems High blood sugar Loss of appetite Vertigo Confusion Memory loss	Wisdom Protection against jealousy (to deflect the evil eye)	SAGE Betony, periwinkle, hops (for vertigo)	Iris, peach, sunflower (for wisdom) Aspen, cumin, garlic, juniper (protection against soul theft through jealousy)
Wou	Ruta graveolens	Headaches Irregular menstruation Tired eyes (as an eye douche) Bruises and sprains (as a compress)	Protects against ill health from negative energies	RUE St. John's wort, agnus castus, skullcap, chamomile, passion-flower (for headache) Arnica, chickweed, hemp, marigold, comfrey (for bruises—as a compress)	Acacia, African violet, bittersweet, cypress, mandrake (for protection)

KREYOL NAME	SCIENTIFIC NAME	MEDICINAL USES	MAGICAL ATTRIBUTES	ANALOGUES & ALTERNATIVES (MEDICINAL)	ANALOGUES & ALTERNATIVES (MAGICAL)
Wowoli	*Sesamum inicum*	Dietary deficiencies Physical and mental weakness Genital and urinary infections	Revelation of deep or hidden knowledge (e.g., "Open sesame!") and powers of prophecy	SESAME (SEEDS) Self-heal (for dietary deficiency) Saw palmetto, damiana (for genito-urinary problems)	Coltsfoot, crocus, kava kava (for wisdom) Pomegranate, violet, walnut (to manifest wishes and the unseen)
Zaboka	*Persea americana*	Stomach ulcers Skin inflammation Burns and scalds Sunburn Respiratory ailments Irregular menstruation	Increases lust and boosts love magic	AVOCADO Comfrey, yarrow, meadowsweet (for peptic ulcers) Lady's mantle, motherwort, primrose (for menstrual problems)	Deer's tongue, radish, southernwood, witch grass (for profound lust)
Zoran'y	*Citrus aurantium*	Poor digestion Loss of appetite Skin inflammation Fungal infections	Enhances the user's beauty/ irresistibility	SWEET ORANGE Aloe, tea tree, daisy, ivy, marigold, poke root (antifungals)	Flax, ginseng, maidenhair (for beauty)

APPENDIX 2

A PERUVIAN HERBAL

Here, we offer selections of frequently used herbs and plantas maestras from Peru, along with their direct Western counterparts or, where necessary, Western analogues for plants that are more easily found. We also list the medicinal uses for these plants as well as their magical properties, so you can continue to explore them on your own.

The first herb listed in column five (in capitals) is the direct counterpart (where there is one) of the Peruvian plant (in column one). Those that are not direct equivalents, but that have similar effects, are listed below it in normal script.

Medical complaints that these herbs address are listed in column three. Analogues and alternatives used in such treatments are given in column five. Specific magical uses and the herbs for these appear in columns four and six.

Again, since the same plants often work for many different conditions, to avoid repetition, not all medical or magical uses are listed for every herb in every case. If you don't see your condition or need against a particular herb, try scanning the other columns and you will almost certainly find it there.

A PERUVIAN HERBAL

PERUVIAN NAME	SCIENTIFIC NAME	MEDICINAL USES	MAGICAL ATTRIBUTES	ANALOGUES AND ALTERNATIVES (MEDICINAL)	ANALOGUES AND ALTERNATIVES (MAGICAL)
Aire sacha	*Kalanchoe pinnata*	Bacterial infections	"The miracle leaf"—Artidoro	BALM	African violet, gardenia, sandalwood (for spiritual strength)
		Viruses		Hypericum, echinacea, garlic (for viruses)	
		Coughs and fevers	Brings inner and outer balance to body and spirit.	Horehound, lungwort (for coughs)	
		Tension	Cures "bad air" (removes negativity and freshens energy)		
		High cholesterol		Garlic (lowers cholesterol)	
		Boils		Chickweed, comfrey, blue flag (for boils—as a compress)	
		Insect bites		Arnica, marigold, myrrh (for insect bites)	
Ajo sacha	*Pseudocalymma alliaceum*	Rheumatism	Plant teacher	"WILD GARLIC"	Peach, sage, sunflower (for wisdom)
		Allergies	Confidence	Echinacea, wild indigo, holy thistle (for rheumatics)	
		Arthritis	Self-knowledge (stalking the self)	Chamomile, centaury, ground ivy (for allergies)	
			Cleansing	Black cohosh, devil's claw, yucca leaves (for osteo-arthritis)	
			Protection		

A PERUVIAN HERBAL (continued)

PERUVIAN NAME	SCIENTIFIC NAME	MEDICINAL USES	MAGICAL ATTRIBUTES	ANALOGUES AND ALTERNATIVES (MEDICINAL)	ANALOGUES AND ALTERNATIVES (MAGICAL)
Albahaca	Ocimum basilicum	Digestive disorders Stomach pain Staphylococcus infection Salmonella Insect bites Alcoholism Poor circulation Depression Deafness	Love Wealth Flying (spirit flight)	BASIL Peppermint, cardamom, fennel (digestives) Chamomile, goldenseal (for staphylococcus infections) Hops, angelica, skullcap (for alcoholism)	Lemon, sweet pea (for friendship) Pecan (to gain employment) Camellia, orange, rice (for wealth) Mugwort, poplar (for astral travel)
Ayahuasca	Banisteriopsis caapi	"Sacred medicine, a master cure for all diseases" —Ralph Metzer All illnesses (when mixed with chacruna and drunk by the shaman, and sometimes the patient, as part of a shamanic healing ritual)	Plant teacher Visions Purgative (cleans toxic energy)	No equivalent	Angelica, coltsfoot, crocus, damiana (for visions) Aloes, butternut, jalap root, senna leaves, yellow dock (spiritual purgatives) Huckleberry, hydrangea, holy thistle, vetivert (spiritual cleansers)

PERUVIAN NAME	SCIENTIFIC NAME	MEDICINAL USES	MAGICAL ATTRIBUTES	ANALOGUES AND ALTERNATIVES (MEDICINAL)	ANALOGUES AND ALTERNATIVES (MAGICAL)
Bobinzana	*Callandra angustifolia*	Rheumatism Postnatal depression Blood disorders Colds and fevers	Plant teacher Spiritual balance Sense of direction in life Self-assuredness	Milk thistle, raspberry leaf (for postnatal depression)	Chamomile, cowslip, wild lettuce (for confidence and balance)
Came	*Peperomia obtusifolia*	Sprains Hernia Burns Sores	Purgative (cleans toxic energy) Energy balancer in the home	BABY RUBBER PLANT Fenugreek, rapturewort (for hernia)	Hawthorn, High John the Conqueror, marjoram, quince (for happiness) Aspen, cumin, juniper (for home protection and harmony)
Castana	*Terminalia catappa*	High cholesterol Dental problems Rheumatism	Wisdom	TROPICAL ALMOND Horsetail, lobelia, cloves (for toothache)	Ginko, ginseng, hawthorn, kola nuts, rosemary (for increased intelligence and enhanced memory)
Chacruna	*Psychotria viridis*	All illnesses (when mixed with ayahuasca and drunk by the shaman, and sometimes the patient, as part of a shamanic healing ritual)	Plant teacher Visionary	No equivalent	See ayahuasca

A PERUVIAN HERBAL (continued)

PERUVIAN NAME	SCIENTIFIC NAME	MEDICINAL USES	MAGICAL ATTRIBUTES	ANALOGUES AND ALTERNATIVES (MEDICINAL)	ANALOGUES AND ALTERNATIVES (MAGICAL)
Chanca piedra	*Phyllantus nirui*	Hepatitis Urinary infections Kidney and gall-stones	Communication at a distance (ESP)	Balmony, barberry, centuary, dandelion, wahoo, wormwood (for hepatics)	Iris, Job's tears, pomegranate, sunflower, violet (enhance ESP)
Chiric sanango	*Brunfelsia grandiflore*	Depression Colds Rheumatism Arthritis	Plant teacher Purifies the home Brings luck Develops the inner self	Passionflower, celery, kola, mugwort, primrose (for depression)	Bay, broom, gum arabic, horseradish, mimosa (for purification)
Coca	*Erythroxylum coca*	Throat and mouth infections Loss of appetite Muscular aches and pains Colic Digestive problems	Divination Opening up the self Offerings to spirit and give-aways of negative energy	Willow, devil's claw, yam, parsley (to ease general aches and pains)	Camphor, cherry, orris (for divination) Grape, rosemary, walnut (to increase mental powers and insight) Dandelion, sweetgrass, tobacco (for calling the spirits)

PERUVIAN NAME	SCIENTIFIC NAME	MEDICINAL USES	MAGICAL ATTRIBUTES	ANALOGUES AND ALTERNATIVES (MEDICINAL)	ANALOGUES AND ALTERNATIVES (MAGICAL)
Comphrey	*Symphytum officinale*	Broken bones Infections Wounds Severe loss of appetite Malnutrition Depression	Financial success (in gambling, new business ventures, etc.) Protection during travel	COMFREY Comfrey, marigold, fenugreek, boneset (for bone injuries) Gotu kola, oats, life root (to prevent cachexia—wasting away)	Feverfew, mugwort (for financial success) Apple, linden (for travel protection)
Herba luisa	*Cymbopogon citratus*	Coughs and chest problems Nervousness Anxiety Poor digestion	Lust	LEMONGRASS Shepherd's purse, bilberry (for severe chesty coughs)	Avocado, endive, mint (to increase lust) Lettuce, vervain, witch hazel (to decrease lust)
Guayusa	*Piper callosum*	Digestive disorders Rheumatism Loss of libido Lack of drive	Lucid dreaming; stalking the self	Gotu kola, ginseng (for increased libido)	Bracken, buchu, jasmine, marigold, rose (for lucid dreaming) Dandelion, dogwood, mandrake (to manifest dreams in ordinary reality)

A PERUVIAN HERBAL (continued)

PERUVIAN NAME	SCIENTIFIC NAME	MEDICINAL USES	MAGICAL ATTRIBUTES	ANALOGUES AND ALTERNATIVES (MEDICINAL)	ANALOGUES AND ALTERNATIVES (MAGICAL)
Jergon sacha	*Dracontium loretense*	Abscesses Hernia Snake bites	Invisibility (against snakes) Spiritual and physical protection during hunts	FER-DE-LANCE Skullcap, yellow dock, burdock (for snake bites)	Chicory, edelweiss, fern, poppy (used in invisibility spells) Mistletoe, primrose, garlic (for hunting success)
Malva	*Malachra ruderalis*	Skin infections Burns Eye irritation Constipation	Heals grief and anger	Senna (to ease constipation)	Clover, balm, motherwort, chamomile (to ease grief) Chamomile, balm (to reduce anger)
Menta	*Menthe piperita*	Anxiety Hypertension Depression	Enhances sleep and dreams Purification	PEPPERMINT Broom, ephedra, gentian (for hypertension)	Agrimony, elder, hops, lavender (for restful sleep) Bracken, jasmine, mimosa (to enhance the power of dreams)
Mucura hembra	*Petiveria alliacea*	Rheumatism Memory loss Headache Loss of energy	Removes bad luck Overcomes fears	Skullcap, gentian, Jamaican dogwood (for loss of energy and "gone all to pieces" syndrome)	Moss, orange, strawberry (to remove bad luck) Loosestrife, myrtle, violet (to remove fears)

PERUVIAN NAME	SCIENTIFIC NAME	MEDICINAL USES	MAGICAL ATTRIBUTES	ANALOGUES AND ALTERNATIVES (MEDICINAL)	ANALOGUES AND ALTERNATIVES (MAGICAL)
Pan del arbol	Artocarpus altilis	Sprains Hernia Bleeding wounds Dental problems	Rootedness, stability Long life	BREADFRUIT Comfrey, arnica (as a poultice for sprains)	Angelica, caraway, castor, holly, pimpernel, willow (for stability) Cypress, maple, tansy (for longevity)
Pinon colorado	Jatropha gossypifolia	Stomach pain Constipation Burns Sprains Intestinal parasites	Calls the healing winds	BELLY-ACHE BUSH Chaparral, garlic, rue (parasiticides)	Broom (calls the wind)
Retama	Cassia reticulata	Liver problems Intestinal parasites Skin diseases	Protection against theft and loss	GOLDEN LANTERN Comfrey, plantain (for liver injuries)	Aspen, garlic, juniper (to protect against theft)
Rosa sisa	Tagetes erecta	Colic Nervous problems Nausea	Soothing to the soul (used in soul retrievals, etc.)	MARIGOLD Valerian, lavender, calamint (for nervous shock) "The seed of calamint relieves infirmities of the heart, taking away melancholy and making a man merry and glad."—John Gerard	Gardenia, meadowsweet, passionflower (for harmony)

A PERUVIAN HERBAL (continued)

PERUVIAN NAME	SCIENTIFIC NAME	MEDICINAL USES	MAGICAL ATTRIBUTES	ANALOGUES AND ALTERNATIVES (MEDICINAL)	ANALOGUES AND ALTERNATIVES (MAGICAL)
Ruda	*Ruta graveolens*	Headaches Blood problems	Luck in business and financial affairs Purifies the energy and the home	Burdock, goldenseal (blood purifiers) Marigold (to aid recovery after blood transfusion)	Buckthorn, hickory (for business and legal success). Coconut (for cleansing the home—by kicking it from room to room, then out into the street)
Santa maria	*Piper peltatum*	Liver problems Conjunctivitis	Spiritual protection for children and the unborn	COW FOOT Eyebright, marshmallow, rose (used as a douche in cases of conjunctivitis)	Nettle, chickweed, dandelion, horsetail, centuary, raspberry leaf, yellow dock (for protection in pregnancy) Gentian, ginseng, chamomile, kelp, marigold (protection for children)
Shapilloja	*Zanthoxylum fragara*	Anxiety Tension Nervousness	Calming and balancing (used in baths) Balances energies	Chamomile, balm, betony (to reduce stress)	Lemon balm, cinnamon, clover, ginger, winter's bark (for balancing and bringing good fortune)

PERUVIAN NAME	SCIENTIFIC NAME	MEDICINAL USES	MAGICAL ATTRIBUTES	ANALOGUES AND ALTERNATIVES (MEDICINAL)	ANALOGUES AND ALTERNATIVES (MAGICAL)
Shimi pampana	*Moranta arundinacae*	Bladder and urinary problems Alcoholism Digestive problems	Energy balancer Brings luck in love	ARROWROOT Cranesbill, shepherd's purse (for bladder problems and enuresis)	Self-heal (balances energy)
Tabaco	*Nicotiana tobacum*	Migraine Nausea Insect bites/ repellent	Plant teacher Spiritual protection Self-knowledge Balances energy Teaches how to heal	TOBACCO Ginger, licorice (to reduce queasiness and general sickness)	Anemone, ash, juniper, knotweed, larkspur, rue, spikenard (for soul healing and protection)
Toe (pronounced "toe-HAY")	*Brugmansia suaveolens*	Ulcers Infections Cancer Abscesses	Plant teacher Seeing into the future Teaches how to heal Removes bad medicine (witchcraft)	ANGEL TRUMPET Cleavers, plantain (to relieve ulcers)	Fig, orange, pomegranate (for divination) Holy thistle, vetivert, wintergreen (to remove curses)
Ucho sanango	*Bonafousia undulata*	Rheumatism Coughs and colds	Plant teacher Visionary Psychic abilities Overcomes fear Removes inner blockages Grounding	Boneset, hyssop (for chesty coughs and colds)	Bladderwrack, borage, eyebright, honeysuckle, star anise, thyme (to increase psychic power)

A PERUVIAN HERBAL (continued)

PERUVIAN NAME	SCIENTIFIC NAME	MEDICINAL USES	MAGICAL ATTRIBUTES	ANALOGUES AND ALTERNATIVES (MEDICINAL)	ANALOGUES AND ALTERNATIVES (MAGICAL)
Una de gato	*Uncaria guianesis*	Joint pains Depressed immune system	Increases sexual enjoyment (especially in women)	CAT'S CLAW Echinacea, lapacho, sage, wild indigo, shiitake mushroom (for autoimmune disease)	Camphor, lettuce (enhances sexual enjoyment) Banana, caper, black cohosh, oak, olive (increases sexual potency in men and women)
Yacu piri piri	*Cyperus sphacelatus*	Snake bites	Luck in love	ROADSIDE FLATSEDGE Onion—bind to the wound (an Irish folk cure for snakebite)	Crocus, High John the Conqueror, lobelia, lotus, moonwort, papaya, tamarind, yohimbe (for luck in love)
Yawar piri piri	*Eleutherine bulbosa*	Fungicide Dysentery Hemorrhaging	Attracts money	Catechu, cranesbill, yam (for amoebic dysentery) Fenugreek, wild indigo, yam (for bacillary dysentery)	Cashew, clover, dock, honesty, jasmine (for money luck)

APPENDIX 3

HOODOO OILS

A few for you to try:

FOR PEACE, HEALING, AND HAPPINESS

"Happy Home" Oil

One way to use this oil is to apply it to candles and burn them in the home for a subtle fragrance that will protect and purify your space. Note that these scents are all citrus based. This corresponds to the notion that citrus fruits such as lemons and limes will cut through negativity and bad magic.

Use 3 drops of orange, 2 each of lemongrass and lemon, and 1 drop of lime added to a base oil.

"Bring Me Peace" Oil

This brings harmony and calm and is especially useful when nervous or upset. It can be used anywhere on the body.

Use 3 drops each of ylang-ylang and lavender, 2 drops of chamomile, and 1 drop of rose added to a base oil.

"Heal Up Fast" Oil

This helps speed the process of healing and can also be applied to bed linen or used in a burner, or diffuser, in the sick room.

Use 4 drops of rosemary, 2 of juniper, and 1 of sandalwood added to a base oil. Alternatively, add the drops to water in a burner.

"Protect Me" Oil

Worn on the body, this offers protection against negative energies and psychic attacks. It can also be used on windows and doors as a guard for the house.

Use 4 drops of basil, 3 of geranium, 2 of pine, and 1 of vertivert added to a base oil.

"Fast Energy" Oil

Wear this to strengthen energy reserves or for a boost when feeling depleted.

Use 4 drops of orange, 2 of lime, and 1 of cardamom added to a base oil.

"Sleep Well" Oil

To bring on natural and restful sleep, use on the temples, neck, the pulse sides of both wrists, and the soles of the feet, or use in a burner in the bedroom.

Use 2 drops of rose, and 1 each of jasmine and chamomile added to a base oil.

FOR LEGAL, FINANCIAL, AND BUSINESS SUCCESS

"Win the Case" Oil

This is used during court cases to confuse the opposition and reveal the holes in their arguments and evidence. Aim to introduce it to the court-room and to attach the fragrance to the opposition but not to yourself. You might, for example, wear it when shaking hands with the opposition counselor before the trial begins, then immediately wash it off your own hands before entering the courtroom.

Use 5 drops of coconut, 3 each of lavender and black pepper, and 2 each of violet and ginseng added to a base oil.

"Bring Me Money Fast" Oil

Wear this on the body and especially on the hands to ensure the fast return of money whenever you touch it, or to gain the support of business partners and influential persons when you shake their hands. You can also use it to anoint green candles, which are burned with the intention of attracting new money.

Use 7 drops of patchouli, 5 of cedarwood, 4 of vertivert, and 2 of ginger added to a base oil.

"Make a Good Impression" Oil

This helps make a favorable impression during interviews and auditions, and will also calm "interview nerves." Wear it as a fragrance.

Use 4 drops of ylang-ylang, 3 of lavender, and 1 of rose added to a base oil.

FOR LUCK

"Fast Luck" Oil

Wear this as a fragrance or dab it onto anything (a photograph, etc.) that represents an area of your life in which you want greater luck.

Use equal parts of cinnamon, vanilla, and wintergreen added to a base oil.

"Reversing the Trick" Oil

To protect against negative energy and return it to whoever sent it, dab this on the pulse points of the wrists.

Use equal parts of lemon, rosemary, rose, and peppermint added to a base oil.

NOTES

Preface

1. Eliot Cowan, *Plant Spirit Medicine: The Healing Power of Plants* (Columbus, N.C.: Granite Publishing, 1991).

Introduction

1. Source: http://ktla.trb.com/news/local/la-me-winesmog, August 22, 2005.
2. Masanobu Fukuoka, *The Road Back to Nature: Regaining the Paradise Lost* (Japan Publications, 1987). Quoted in Stephen Harrod Buhner, *The Secret Teachings of Plants: The Intelligence of the Heart in the Direct Perception of Nature* (Rochester, Vt.: Bear & Company, 2004).

Chapter 1. Nothing Is Hidden: How Plants Heal

1. See, for example, Nicholas Goodrick-Clarke, *Paracelsus: Essential Readings* (Berkeley, Calif.: North Atlantic Books, 1999).
2. Thomas Bartram, *Bartram's Encyclopedia of Herbal Medicine: The Definitive Guide to the Herbal Treatment of Diseases* (London: Robinson Publishing, 1998).
3. See, for example, James Lovelock, *Gaia: A New Look at Life on Earth* (London: Oxford University Press, 2000).
4. James Lovelock, "The Earth Is About To Catch A Morbid Fever That May Last As Long As 100,000 Years," London, *The Independent*, 16 January 2006.

5. Sir James Frazer, *The Golden Bough: The Roots of Religion and Folklore* (London: Penguin Books, 1996). This book was originally published in 1890 in two volumes as *The Golden Bough: A Study in Comparative Religion.*

6. Frank J. Lipp, PhD, *Herbalism: Healing and Harmony, Symbolism, Ritual and Folklore Traditions of East and West* (London: Duncan Baird Publishers, 1996).

7. Ibid.

8. Michele Peterson, "Bewitched in Bolivia," London, *The Globe and Mail*, Saturday, 30 October 2004, p. T7. http://www.theglobeandmail.com/serv let/ArticleNews/TPStory/LAC/20041030/BOLIVIA30/TPTravel/

9. Lovelock, "The Earth Is About To Catch A Morbid Fever That May Last As Long As 100,000 Years."

10. Marie-Louise Von Franz, *The Interpretation of Fairy Tales* (Boston: Shambhala, 1996).

11. Peter Tompkins and Christopher Bird, *The Secret Life of Plants* (New York: HarperCollins, 1973).

12. Ibid.

13. Ibid.

14. In Kay C. Whittaker, *The Reluctant Shaman: A Woman's First Encounters with the Unseen Spirits of the Earth* (San Francisco: HarperSanFrancisco, 1991).

15. In Tompkins and Bird, *The Secret Life of Plants.*

16. Alfred Vogel, *The Nature Doctor: A Manual of Traditional and Complementary Medicine* (New Canaan, Conn.: Keats Publishing, 1991).

17. Jeremy Narby, *Intelligence in Nature: An Inquiry into Knowledge* (New York: Jeremy P. Tarcher, 2005).

Chapter 2. The Shaman's Diet: Listening to the Plants

1. Richard E. Schultes and Michael Winkelman, "The Principle American Hallucinogenic Plants and their Bioactive and Therapeutic Properties." In Michael Winkelman and Walter Andritzky, eds., *Yearbook of Cross-Cultural Medicine and Psychotherapy* (VWM—Verlag fur Wissenschaft und Bildung, 1995).

2. John Michell, *Confessions of a Radical Traditionalist* (Waterbury Center, Vt.: Dominion Press, 2005).

3. In John G. Neihardt, *Black Elk Speaks: Being the Life Story of a Holy Man of the Oglala Sioux* (Lincoln: University of Nebraska Press, 1932).

4. Michael Harner, *The Jivaro: People of the Sacred Waterfalls* (Berkeley: University of California Press, 1984).

5. Maria Sabina, *Maria Sabina and Her Mazatec Mushroom Velada* (New York: Harcourt Brace Jovanovich, 1974).

6. Thomas Bartram, *Bartram's Encyclopedia of Herbal Medicine.*

7. For more on Rivas and his work, see Jaya Bear, *Amazon Magic: The Life Story of Ayahuasquero and Shaman Don Agustin Rivas Vasquez* (El Prado, New Mexico: Colibri Publishing, 2000).

Chapter 3. Plants of Vision: Sacred Hallucinogens

1. See, for example, Terence McKenna, *Food of the Gods: The Search for the Original Tree of Knowledge, A Radical History of Plants, Drugs, and Human Evolution* (New York: Bantam Books, 1993).

2. Rita Carter, *Mapping the Mind* (Berkeley: University of California Press, 1998).

3. Jeremy Narby, *The Cosmic Serpent* (New York: Jeremy P. Tarcher, 1999).

4. Joseph Campbell, *Pathways to Bliss: Mythology and Personal Transformation* (Novato, Calif.: New World Library, 2004).

5. Terence McKenna, *Food of the Gods.*

6. Richard Evans Schultes, *Plants of the Gods: Their Sacred, Healing, and Hallucinogenic Powers* (Rochester, Vt.: Healing Arts Press, 2001).

7. Ibid.

8. Bill Hicks, in *Love All The People: Letters, Lyrics, Routines,* edited by John Lahr (London: Constable & Robinson, 2004).

9. A good introduction to Lilly's work is his book, *The Center of the Cyclone* (New York: Three Rivers Press, 1985). A practical analysis of his work also appears in: Ross Heaven, *The Journey To You* (New York: Bantam Press, 2001), and for further exploration of sensory deprivation and the shamanic effects of time spent in darkness, see: Ross Heaven, *Darkness Visible* (Rochester, Vt.: Destiny Books, 2005).

10. Katy Weitz, "Stressed out? Put on a blindfold for 72 hours," London, *The Observer,* 5 June 2005.

Chapter 4. Healing the Soul

1. Malidoma Somé, *Of Water and the Spirit: Ritual, Magic, and Initiation in the Life of an African Shaman* (New York: Penguin Books, 1995).

2. Martin Prechtel, *Secrets of the Talking Jaguar: Memoirs from the Living Heart of a Mayan Village* (New York: Jeremy P. Tarcher, 1999).

3. Michael Harner, *The Way of the Shaman* (San Francisco: HarperSanFrancisco, 1990).

4. Sandra Ingerman, *Soul Retrieval: Mending the Fragmented Self Through Shamanic Practice* (San Francisco: HarperSanFrancisco, 1991).

5. Ibid.

6. Stanley Krippner and Alberto Villoldo, *The Realms of Healing* (Berkeley, Calif: Celestial Arts, 1987).

7. See *Spirit in the City* by Ross Heaven (New York: Bantam Press, 2002).

8. Maya Deren, *Divine Horsemen: The Living Gods of Haiti* (New York: McPherson, 1983).

9. Joseph Campbell, *The Masks of God,* from the *Primitive Mythology* volume (New York, Penguin Books, 1991).

10. Carl G. Jung, *Mysterium Conjunctionis* (Paris: Albin Michel, 1989).

11. Originally published in *Pravda*, March 12, 2005. Reported in "A Hearty Afterlife," *Fortean Times* magazine, issue 197, June, 2005.

12. *Mexican Teachings: Plant Spirits in Ceremony,* author unknown, at the Web site www.dpw.wageningen-ur.nl.

13. In Thomas Bartram, Bartram's *Encyclopedia of Herbal Medicine.*

Chapter 5. Pusangas and Perfumes: Aromas for Love and Wholeness

1. Ilza Veith, trans., *The Yellow Emperor's Classic of Internal Medicine* (Berkeley: University of California Press, 2002).

2. David Hoffman, *Welsh Herbal Medicine* (Cardigan, Ceredigion, Wales: Abercastle Publications, 1978).

3. *The Practice of Aromatherapy* was reprinted by Healing Arts Press, Rochester, Vt., in 1982.

Chapter 6. Floral Baths: Bathing in Nature's Riches

1. David Hoffmann, *Welsh Herbal Medicine.*

2. Nadine Epstein, "A Maya Spiritual Bath that Heals a Child's Nightmares." *Mothering* magazine, 1989. Online at www.mothering.com/articles/new _baby/sleep/epstein.

3. Sir James Frazer, *The Golden Bough.*

4. Michio Kaku, *Hyperspace: A Scientific Odyssey through Parallel Universes, Time Warps, and the 10th Dimension* (New York: Anchor Books, 1995).

5. Jeremy Narby, PhD, *Intelligence in Nature: An Inquiry into Knowledge* (New York: Jeremy P. Tarcher, 2005).

6. Ibid.

7. Morwyn, *Magic from Brazil: Recipes, Spells and Rituals* (St. Paul, Minn.: Llewellyn Publications, 2001).

8. Ibid.

9. Viktor Frankl, *Man's Search for Meaning: The Classic Tribute to Hope from the Holocaust* (London: Rider/Random House, 1992).

10. Max Beauvoir, writing on the Web site http://www.vodou.org/treatmen1 .htm.

11. Ibid.

12. *Cambridge Relativity* Web site: http://www.damtp.cam.ac.uk/user/gr/ public/index.html.

13. Ibid.

14. Fred Alan Wolf, PhD, *The Dreaming Universe* (New York: Touchstone, 1995).

15. Zora Neale Hurston, *Tell My Horse* (New York: Harper and Row, 1996).

Chapter 7. The Scream of the Mandrake

1. Patrick Logan, *Irish Folk Medicine* (London: Appletree Press, 1999).

2. Ibid.

3. Jeremy Laurance, "Why This Exotic Fruit Is the World's Only Weapon against Bird Flu," London, *The Independent,* 15 October 2005.

4. Kevin Trudeau, *Natural Cures "They" Don't Want You to Know About* (Worcester, U.K.: Alliance Publishing, 2005).

5. Jerome Burne, "Turn Over a New Leaf," London, *The Independent,* 10 May 2005.

6. As reported in a news story at www.yahoo.com, July 2005.

7. Jerome Burne, "Turn Over a New Leaf."

8. Ibid.

9. Luis Eduardo Luna and Pablo Amaringo, *Ayahuasca Visions: The Religious Iconography of a Peruvian Shaman* (Berkeley, Calif.: North Atlantic Books, 1993).

10. Ibid.

11. Patrick Logan, *Irish Folk Medicine.*

GLOSSARY

This section provides basic definitions and explanations for some of the more common shamanic terms used throughout the text. It does not include plant names or explanations, as these are covered in the text itself and in appendices 1 and 2. For the sake of brevity and to avoid repetition, it also excludes less frequently used terms and those that are extensively explained in the relevant chapters.

artes: Medicine tools used by San Pedro shamans, which are kept on a mesa (altar) during healing ceremonies and may include statutes of the saints, herbs, candles, swords, and so on, all of which are used to cleanse the energy field of a patient.

ayahuascero, ayahuascera: An Amazonian plant shaman who works with ayahuasca, the visionary "vine of souls."

banjos florales: *See* floral baths.

brujo: A shaman or sorcerer who is regarded as practicing "evil" or "black" magic.

chacapa: A medicine rattle made from dried leaves, used by ayahuasca shamans.

chantes: In Haiti, songs that are calls to particular healing spirits.

chungana: The rattle used by a San Pedro shaman.

curandero, curandera: In Spanish-speaking countries, a traditional healer who may also work with plants.

diet: The practice of following certain prescriptions and prohibitions for a defined period of time in order to open oneself spiritually to the plants. The diet involves taking (and avoiding) certain foodstuffs, but there are also restrictions on activities such as sexual intercourse and mixing with other people.

divination: A traditional method of patient diagnosis or healing that usually employs plants, such as coca (Andean shamanism) or nuts (Ifá) to understand an illness in its widest spiritual sense.

doctrine of signatures: The idea, developed by the physician Paracelsus, that the Creator has left his mark on plants so they reveal their healing properties through their outward appearance, or signatures.

dominio: The process of transferring power to a patient through magical means.

energy body: The luminous egg of energy that surrounds and infuses a human being. Often this is divided into bands of light, which refer to the spiritual, emotional, mental, and physical selves contained within this egg. In traditional societies, the energy body may simply be called the soul.

envidia: Envy or jealousy. In Peru and Haiti, envy is a magical force that can cause harm to a person through spirit intrusion or the "evil eye." *Also see* mal d'ojo.

floral baths: Sacred baths prepared and delivered by shamans to revitalize and rebalance the energy body or spirit.

flourishing: The shamanic process of *intending* success and well-being for a patient. This may be carried out by spraying perfumes over the patient (known in Peru as *camaying*).

folk medicine: The traditional healing practices of Europe, which almost always involve plant cures and may also include magical practices such as incantations, chants, prayers, and bathing in holy wells.

gris-gris: In New Orleans Hoodoo, a medicine bag containing plants and magical items such as charms and crystals, which is charged with a particular intention (e.g., to attract luck, love, or money to its owner). Also known as a mojo bag.

icaros: Magical songs of power that are taught to shamans by their plant spirit allies and that bring healing and guidance to a patient. *Also see* chantes.

limpia: A cleansing, usually given by running a plant, egg, or other object such as a stone through a patient's energy field in order to remove spirit intrusions and restore healing and balance.

Lwa (or Loa): In Haiti, angel-like spirits, regarded as aspects of God. There are more than five hundred different Lwa, all of whom have their own personalities and healing skills and are appealed to in different and unique ways.

macerado: A plant medicine made by macerating leaves in alcohol. *Also see* tincture.

mal d'ojo: The "evil eye." A way of sending negative energy toward someone, usually prompted by jealousy. *Also see* envidia.

medsen fey: In Haiti, a "leaf doctor" or herbalist who works with the Lwa and the spirit of the plants.

mesa: A "high place." In the Andes, the altar used by a San Pedro shaman. *Also see* artes.

mojo: *See* gris-gris.

moraya: In the Shipibo tradition, the highest level of plant shaman, requiring years of experience and special dieting. *Also see* diet.

offerenda: An offering made to the spirits of nature, often as part of a healing.

otherworld: The world of spirit or nature that shamans explore on their journeys. The otherworld is the home of helpful spirits and a place where soul parts can also take refuge. *Also see* soul loss and soul retrieval.

paket: A Haitian medicine tool. Pakets are highly decorated bags of magical herbs that have the dual intention of removing negative energies from a patient and giving healing in return.

perfumero, perfumera: A Peruvian shaman who specializes in the use of fragrances for affecting material change. Skilled perfumeros can use fragrance to bring about healing, to increase business success, enhance fortunes in love, help a patient win a legal battle, and so forth.

plantas maestras: Master teachers in plant spirit form, such as ayahuasca and San Pedro.

platicas: In Aztec and Andean cultures, a form of therapeutic consultation and counseling that is carried out heart-to-heart and soul-to-soul between healer and patient.

pusanga: The "love medicine of the Amazon." A special plant perfume with the power to attract good fortune and human admirers (for the purpose of making them fall hopelessly in love with the person wearing the perfume, etc.).

saladera: A run of bad luck, usually caused by energy imbalances, which can be corrected by realigning the patient with natural forces and reconnecting him or her to the Earth.

seguros: Plant amulets or charms that draw good fortune in a number of ways according to the plants that are used. In the Andes, these charms become friends and are also used as confidants in therapy-based self-healing sessions.

sin: In Welsh sin-eating practice, a "weight on the soul" (or spirit intrusion) caused by guilt or shame, or by carrying the guilt of another. It can be removed through confession and atonement, or by the intervention of a sin eater who will perform a spirit extraction.

sin eater: A Celtic (Welsh, Irish, Scottish) shaman who is expert at spirit extraction on the living and the dead, and at healing with herbs and plants. .

singado: An Andean technique of energy rebalancing and spiritual cleansing, where tobacco macerated in rum and honey is snorted into the nostrils (left nostril to release negativity, right to draw in benign forces).

soul loss/retrieval: Soul loss is a spiritual disease caused by shock or trauma, where the soul fragments and part of its energy is lost. Soul retrieval is the process of finding and returning these missing parts. This is accomplished by the shaman journeying to the spirit world and carrying the soul pieces (or energy) back to the patient and/or by offerings made to nature. *Also see* offerenda.

spirit intrusion/extraction: A spirit intrusion is an infection of energy that is unhelpful or debilitating to a patient, often caused by explosions of anger or sent by a rival as a result of envidia (envy). Extraction is the process of removing this energy, most often through spirit negotiation followed by the use of plant medicines.

tea: A simple infusion of herbs or plants in hot water. Honey may be added to taste.

tincture: In modern herbalism, concentrated liquid extracts of soluble plant constituents in 25 percent ethyl alcohol. In traditional plant shamanism, plant allies kept in a rum or vodka base. *Also see* macerado.

tonic: Plant medicines that impart strength and vitality.

INDEX

CP denotes color plates.

ABOUT THE AUTHORS

Ross Heaven is the founder and director of The Four Gates Foundation, one of Europe's leading organizations for the teaching, promotion, and application of spiritual wisdom and freedom psychology. Ross offers workshops in healing, empowerment, Plant Spirit Shamanism, and indigenous wisdom. He is also a Western-trained therapist and the author of eight other books on shamanism, healing, and personal development, including *The Journey To You, Spirit In the City, Vodou Shaman, Darkness Visible,* and *The Spiritual Practices of the Ninja,* as well as *Infinite Journeys,* a trance drumming tape that can be used as an accompaniment to the shamanic journeys described in these books. Ross has worked extensively with the shamans of the Amazon and Andes, North America, and Europe and has been initiated into the medicine traditions of Haiti and the sin-eating practices and plant healing traditions of Celtic Britain. He runs trips overseas to work with the indigenous healers from these traditions and offers Plant Spirit Shamanism courses and medicine retreats in the UK, Ireland, France, Haiti, the United States, and the Amazon basin, where participants are able to experience the direct healing of nature and explore their vision and creativity with ayahuasca and other teacher plants. For details of Plant Spirit Shamanism courses, as well as book extracts, articles, and other workshops with Ross Heaven see his Web site: www.thefourgates.com.

Howard G. Charing is a healer, a workshop facilitator, and a director of the Eagle's Wing Centre for Contemporary Shamanism. He has written numerous articles on Amazonian plant medicines and has worked with some of the leading shamans in this area. Howard has also produced a collection of traditional chants and icaros on a CD entitled *The Shamans of Peru*. For many years he has been organizing plant medicine retreats to work with the Shipibo Indians in Peru's Amazon basin, as well as Andean retreats with the San Pedro shamans of northern Peru. Howard has been baptized by the Shipibo and ritually initiated into the lineage of the maestros of the Rio Napa community. Howard and his colleague Peter Cloudsley also hold Amazonian medicine retreats at their dedicated Center in the Mishana National Park. As a protected nature reserve encompassing several thousand acres of Peruvian jungle, Mishana is home to many plant and animal species that are not found anywhere else on the planet. At Mishana, participants are able to work with indigenous shamans and experience the direct healing of nature through the visionary vine of souls. For details about Peruvian plant medicine retreats, as well as information about upcoming workshops with Howard G. Charing see his Web site: www.shamanism.co.uk.